Mansfield Merriman

Elements of sanitary Engineering

Mansfield Merriman

Elements of sanitary Engineering

ISBN/EAN: 9783743343436

Manufactured in Europe, USA, Canada, Australia, Japa

Cover: Foto ©ninafisch / pixelio.de

Manufactured and distributed by brebook publishing software (www.brebook.com)

Mansfield Merriman

Elements of sanitary Engineering

ELEMENTS

OF

SANITARY ENGINEERING.

BY

MANSFIELD MERRIMAN,
PROFESSOR OF CIVIL ENGINEERING IN LEHIGH UNIVERSITY.

SECOND EDITION.

FIRST THOUSAND.

NEW YORK:
JOHN WILEY & SONS.
LONDON: CHAPMAN & HALL, LIMITED.
1899.

Copyright, 1898,
BY
MANSFIELD MERRIMAN.

ROBERT DRUMMOND, PRINTER, NEW YORK.

CONTENTS.

CHAPTER I.

SANITARY SCIENCE.

			PAGE
ART.	1.	INTRODUCTION.	7
	2.	HISTORICAL NOTES.	9
	3.	CLASSIFICATION OF DISEASES.	12
	4.	STATISTICS OF MORTALITY	15
	5.	BACTERIOLOGY.	18
	6.	ORGANIC MATTER.	21
	7.	FILTH AND DISEASE	23
	8.	IMPURE AIR AND DISEASE	25
	9.	DRINKING WATER AND DISEASE	28
	10.	MATTER IN NATURAL WATERS	30
	11.	CHEMICAL ANALYSIS OF WATER	33
	12.	BIOLOGICAL ANALYSIS OF WATER	36
	13.	INTERPRETATION OF ANALYSES	39
	14.	RESULTS OF SANITARY SCIENCE	42
	15.	EXERCISES AND PROBLEMS	44

CHAPTER II.

WATER AND ITS PURIFICATION.

ART.	16.	THE RAINFALL	47
	17.	EVAPORATION, RUN-OFF, AND PERCOLATION.	51
	18.	RAIN WATER.	52
	19.	SURFACE WATERS.	54
	20.	GROUND WATERS.	57
	21.	RESERVOIRS	60

		PAGE
Art. 22.	Sedimentation and Aeration	63
23.	Natural Filtration	66
24.	Artificial Methods of Purification	68
25.	Chemical and Electrical Methods	70
26.	Screens and Strainers	72
27.	Mechanical Filters	74
28.	Artificial Filtration	77
29.	Operation of Filter Beds	80
30.	Exercises and Problems	84

Chapter III.

WATER-SUPPLY SYSTEMS.

Art. 31.	Classification	86
32.	Consumption of Water	89
33.	Capacity of Storage Reservoirs	91
34.	Reservoir Dams of Earth	94
35.	Reservoir Dams of Masonry	97
36.	Waste-weirs and Pipe Connections	101
37.	Aqueducts	103
38.	Pipe Lines	107
39.	Distributing Reservoirs	110
40.	Pumps and Pumping	113
41.	Pumping Engines	116
42.	Pumping to Reservoirs	120
43.	Direct Pumping	122
44.	Tanks and Stand Pipes	126
45.	Street Mains and Fire Service	129
46.	Water Meters and House Pipes	132
47.	Exercises and Problems	135

Chapter IV.

SEWERAGE SYSTEMS.

Art. 48.	Historical Notes	139
49.	House Fixtures	143
50.	House Drainage	146

CONTENTS.

		PAGE
ART. 51.	CLASSIFICATION OF SYSTEMS	149
52.	THE COMBINED SYSTEM	152
53.	THE SEPARATE SYSTEM	156
54.	SIZES OF SEWERS	159
55.	CONSTRUCTION OF SEWERS	162
56.	VENTILATION AND CLEANING	165
57.	PUMPING OF SEWAGE	168
58.	VACUUM SYSTEMS	170
59.	THE COMPRESSED-AIR SYSTEM	173
60.	COST AND ASSESSMENTS	175
61.	EXERCISES AND PROBLEMS	178

CHAPTER V.

DISPOSAL OF GARBAGE AND SEWAGE.

		PAGE
ART. 62.	STREET CLEANING	180
63.	REMOVAL OF GARBAGE	183
64.	CREMATION OF GARBAGE	186
65.	DIGESTION OF GARBAGE	188
66.	CLEANING OF CESSPOOLS	190
67.	SEWAGE AND ITS DECOMPOSITION	192
68.	DISPOSAL OF SEWAGE IN RIVERS	195
69.	SCREENING AND AERATION	197
70.	CHEMICAL PRECIPITATION	200
71.	INTERMITTENT FILTRATION	205
72.	BROAD IRRIGATION	209
73.	CONCLUSION	213
74.	EXERCISES AND PROBLEMS	215

INDEX 217

ELEMENTS OF SANITARY ENGINEERING.

Chapter I.

SANITARY SCIENCE.

1. Introduction.

Sanitary science embraces those principles and methods by which the health of a community is promoted and the spread of disease is prevented. Hygiene properly relates to the individual or to the family, but sanitary science has a wider scope and includes the village, the city, and the community at large. Hygiene is the preservation of the health of the individual under the rules of the physician, while sanitary science has for its aim the preservation and protection of the health of the community under the combined action of physicians, engineers, and the civil authorities.

The field of sanitary science is a wide one. It includes the collection of vital statistics, and particularly the statistics of mortality and disease, the isolation and quarantine of infectious diseases, the disinfection of houses, the management of hospitals, and the proper burial of the dead. It embraces all the regulations for preventing adulteration of food and pollution of air or water. It treats of the methods of heating and ventilating public buildings so as to promote comfort and health, of the methods for securing pure and abundant supplies of water, of the drainage of land, and of the removal of garbage and sewage. To properly coordinate all these subjects, sanitary science uses the principles of biology, chemistry, medicine, physics, and engineering, and from these it frames regulations to be enforced by the civil authorities. It invokes the science of the biologist and chemist, the experience of the physician, the constructive talent of the engineer, and the authority of the legislature in order to preserve and protect the health of the community.

In this chapter only a small part of the field of sanitary science can be discussed, and the portions to be selected are those which are most directly applicable to the work of the sanitary engineer. The engineer cannot be a biologist or a chemist, but he must be able to understand the main reasons for their conclusions. He cannot be a physician, but he should know something about the general subject of the prevention of disease. He cannot be skilled in architecture, but he should understand the fundamental principles relating to heating and ventilating. He cannot be an expert in social science or in law, but he should not be ignorant of the methods by which vital statistics are collected and sanitary regulations are enforced. This chapter aims to briefly explain some of these subjects, in order that the student may obtain a broad view of the whole field, and that the engineer may be better able to effectively cooperate with the other professions in advancing the sanitary condition of the community.

Civil Engineering is the art of economic construction, that is, the art of making structures for the public use at the minimum cost for installation and operation. A person not an engineer may construct a railroad or a water-supply system, but its cost will be higher and its efficiency lower than one built by the engineer of experience. The engineer, according to the definition of Telford, utilizes the materials and forces of nature for the benefit of man, but to this should be added that in so doing he aims to secure the least possible cost of construction and maintenance.

Sanitary engineering is that branch of civil engineering which is concerned with the constructions for promoting the health of the community. Such constructions fall into two classes, called Water Supply and Sewerage, and these form the subjects for the following chapters. A pure and abundant water supply, and an efficient system of sewerage, have been universally found to promote cleanliness and prevent the spread of disease; to construct these in an economical manner is the main work of the sanitary engineer.

2. Historical Notes.

The oldest sanitary code on record is that given in the book of Leviticus for the guidance of the Israelites while traveling in the Arabian deserts; the date of this code is about 1490 B.C. It includes rules regarding the kinds of meat to be eaten, the degrees of consanguinity within which marriages were forbidden, the inspection and isolation of unclean or diseased persons, and particularly of lepers, but it gives no directions regarding the purity of the water supply or the removal of garbage and refuse. Nevertheless, it is thought by many that the strict observance of this sanitary code has been one of the causes of the remarkable vitality of the Jewish race.

The Moabite stone which records the rebellion of Mesha, king of Moab, against the Israelites, about 900 B.C., states that Mesha built two conduits and ordered the inhabitants of the city of Karcha to place a cistern in each house. Prior to this the Egyptians had built reservoirs and canals for the purpose of irrigation. Nebuchadnezzar also constructed reservoirs and canals near Babylon about 590 B.C. Jerusalem had a water supply furnished by subterranean aqueducts, some of which are yet in operation.

The Romans at first built canals for drainage and irrigation, and later constructed extensive systems of aqueducts for the water supply of Rome and other cities. The aqueducts of Rome in 97 A.D. were described by Frontinus, a military engineer, who was then the imperial water commissioner; these included 30 miles of conduits built on arches and 220 miles under the surface of the ground, and later there were fourteen aqueducts having an aggregate length of 359 miles. It has been frequently stated that these aqueducts delivered to Rome about 300 million gallons of water per day, or, since the population of the city was nearly one million, about 300 gallons per person per day. A recent discussion by Herschel indicates, however, that this figure is much too high, and that probably the average daily supply within the city was not far from 50 American gallons per person. Even this quantity was a liberal supply for domestic and public use.

After the division of the Roman empire, about 300 A.D., and the subsequent conquests by the barbarians, the splendid roads and aqueducts were suffered to fall into decay, and Europe for more than a thousand years was pervaded by intellectual and social darkness. Unfortunately the teachings and practice of the Christian church during this period regarded cleanliness as one of the luxuries which was inconsistent with godliness, while bodily filth was considered as a mark of inward piety and holy sanctification. The example set by the monastic orders was imitated by the people at large; bathing was unknown, houses and clothing were filthy, and the streets served as receptacles for garbage and human excreta. The result of this violation of sanitary principles is seen in the horrible pestilences that spread over Europe. One of these, called the Black Death, is said to have had 40 000 000 victims since its first appearance in the fourteenth century. At this time the common people lived in a manner more revolting than that of barbarous tribes at the present day; oppressed by the lords and by the clergy, they had few civil rights, no liberty of thought, and their minds were actuated by the grossest superstitions.

The revival of learning, the invention of printing, and the discovery of America characterized the close of the fifteenth century and led the way for the religious reformation of the sixteenth. Then ensued that struggle for liberty of thought and freedom of conscience which is not yet entirely ended. Later the study of the laws of nature began slowly to banish superstition and to show that disease and pestilence were not punishments arbitrarily sent from heaven, but that they resulted from man's own neglect of sanitary principles. Harvey in 1619 discovered the facts regarding the circulation of blood in the veins and arteries, but it was not until after the discovery of oxygen by Priestley, in 1774, that the phenomena could be fully understood. Medical science slowly outgrew its superstitions and began to found its methods of treating disease upon observed facts rather than upon arbitrary fancies. How slow the progress had been from the fifteenth to the close of the eighteenth century may be judged from the fact that when Jenner, in 1798, announced his discovery of vaccination as a preventive of smallpox both physicians and clergy almost unanimously denounced and derided him in the strongest language.

During the first quarter of the nineteenth century the essential elements of railroad locomotion were perfected, thereby greatly increasing traffic and commerce, and an era of material progress was inaugurated which was highly conducive to the development of science. Physics, chemistry, and biology began to be studied from a new point of view, and all the various subjects which are now embraced in the art of engineering began to receive zealous attention. Road improvements, the drainage of towns, and the construction of water supplies began. Drains to carry off the rainfall had been built in London during the seventeenth century, but it was not until 1815 that they were allowed to be used to carry away sewage; before the middle of the century, however, a great change of opinion had occurred, and in 1847 it was made compulsory to turn all sewage into such drains or sewers. At this date modern sanitary engineering may be said to have had its origin, although it was not until many years later that the term became known to the public. The progress that has been made since that time will be recorded in the following pages under the several subdivisions of the subject.

3. Classification of Diseases.

The ancient idea regarding disease seems to have been that health was the normal state of man, and that disease was a punishment inflicted by an arbitrary divine authority. Later this idea became modified so as to regard disease as the result of the violation of sanitary laws, health being still considered as the normal condition. This point of view is still not uncommon, and such terms as " preservation of health " and " protection from disease " tend to imply that it is a correct one. Strictly speaking, however, health is an ideal state, but not a normal state. The normal or natural conditions of life on this earth are those of disease. We are born with constitutional weaknesses inherited from our ancestors; our food, water, and air contain poisons; extremes of cold and heat prevent ideal growth; infectious germs and injurious insects are continually making attacks on us; in short, life is a struggle for existence. These influences are particularly powerful in childhood, and statistics show that over one-fourth of all deaths are those of infants under one year of age. Of all animals, man is the most helpless at birth, and it is safe to say that without the watchful care of parents or friends very few children could survive for one year, even if suitable food were at hand.

It therefore appears that health should not be regarded as the normal state of man under natural conditions, but rather as an ideal state which might occur under ideal conditions. Disease, in strictness, is to be regarded as the normal state.

CLASSIFICATION OF DISEASES.

and health as the state insured by eternal vigilance in removing the causes that continually tend to produce disease and death.

Disease in general may be defined as a derangement of the organs or tissues of the body whereby their functions cannot be properly performed. Deaths by violence or accident, those of women in childbirth, those of children due to teething, and those resulting from starvation, excessive labor, and old age need not here be considered. Local diseases, such as those of the brain and heart, and those of the digestive, circulatory and generative systems, together with constitutional diseases like rheumatism, cancer, and scrofula, constitute a large proportion of all ailments, but their prevention is the province of the physician under the rules of hygiene, and these may need to be very different for different individuals.

Zymotic diseases are those caused by infection from without the individual, and they usually have a period of incubation, followed by illness with fever and perhaps accompanied by eruption. Among these are smallpox, measles, diphtheria, whooping-cough, typhoid fever, influenza, intermittent fever or ague, diarrhœa, cholera infantum, cholera, and yellow fever. Syphilis, gangrene, and hydrophobia, although allied to the zymotic class, are generally called enthetic or inoculated diseases. Tuberculosis, or consumption, was formerly regarded as a local constitutional disease, but it is now recognized as zymotic in the sense that it is caused by infection, hereditary tendencies merely rendering its development more rapid in some individuals than in others. It is the province of sanitary science to abolish these diseases by removing their causes.

All zymotic diseases are produced or propagated by organic germs or microbes. For instance, malaria is produced by a certain germ whose home is in swamps or low lands; typhoid fever is caused by a germ which lives in certain fouled waters; diphtheria, measles, scarlet fever, smallpox, yellow fever, and cholera have their own especial germs, which are communicated from one person to another through air or water. Influenza, or grip, is supposed to be caused by a peculiar germ which lives in fine volcanic dust, and the last epidemic of this may perhaps be due to the great eruption at Krakatoa in 1883. If it were possible to utterly annihilate all such living germs these diseases would entirely disappear.

Zymotic or germ diseases are called contagious or infectious, the first word implying touch or contact, while the second does not, although the words are frequently used without such distinction. A contagious disease may not be infectious, but an infectious disease is generally also communicated by contagion; for instance, syphilis is contagious but not infectious; on the other hand, smallpox has been known to cross a river a mile in width, and hence may be called infectious, while at the same time its contagious power is far greater.

An endemic disease is one peculiar to a certain locality, or one which appears regularly at such localities. For instance, in certain mountains of Europe large numbers of people are afflicted with goitre; in some low lands malaria is always found; again, at some tropical seaports yellow fever may be expected to appear regularly in certain months of the year.

An epidemic disease is one that spreads over a community at irregular intervals and then for the most part disappears. Smallpox was formerly one of the best examples of an epidemic disease, but it is now kept under control by vaccination and isolation, so that few epidemics occur in civilized countries. Measles, whooping-cough, and other diseases of childhood often become epidemic in a town or city. Typhoid fever is one of the most dangerous of the epidemic diseases, its spread being wholly due to germs which have infected the water supply. Diphtheria is epidemic in neighborhoods, but not generally over large areas. The prevention of dangerous epidemics of zymotic diseases has been one of the great sanitary triumphs of the nineteenth century. Epidemics of smallpox, yellow fever, and cholera caused great devastation in America a hundred years ago; now they are under full control and their spread is no longer feared.

4. Statistics of Mortality.

Statistics of births, marriages, and deaths are absolutely necessary for the study of social and economic science. The record of deaths, with statement of the causes of death, is equally necessary for the successful progress of sanitary science. All large cities, many towns, and some states now require such a registration, while medical societies and hospitals are doing most excellent work in collecting records of illness and recovery, as well as of death. The attempt to collect mortality statistics at the decennial census results in apparent failure, on account of the omission of many cases by the enumerators, but even the facts thus collected are of value in comparative studies, and fortunately these may be rendered more reliable by being adjusted to correspond with the records of the registration of cities and states. New Hampshire, Vermont, Massachusetts, Rhode Island, Connecticut, New York, New Jersey, Delaware, and Alabama are the only American states that had satisfactory systems of registration in 1890, and a comparison of their records with those of the census enumerators showed that the latter failed to report about 30 per cent of the deaths.

To obtain correct mortality statistics it is necessary that the law should require, under a heavy penalty, that no undertaker or other person shall inter, or remove from the town, the body of a deceased person without a permit from the board of health; that such permit shall not be issued until a physician has filed a certificate with the board of health giving the name, sex, age, color, nature of illness, and cause of death of that person. In this manner every death in every town immediately goes on record, and the local boards of health transmit these records monthly to the state authorities. The attempt to collect this information annually through assessors, as has been done, in Pennsylvania, results in entire failure and in the waste of public money. The method above described is simple and effectively collects the desired facts. Moreover, as the physicians' certificates are daily placed on file, the board of health has constant information as to the degree of prevalence of each disease in the town and is able to take proper measures to prevent the spread of infection.

The average yearly rate of mortality in the United States is now about 18 deaths for every thousand inhabitants. In the cities the yearly rate is about 23 deaths per thousand, while in the rural districts it is about 15. In 1890 the highest rate for the white population of any city was 29.0 for Savannah, Ga., while the lowest was 14.4 for Nashville, Tenn. In cities the mortality of the colored population is materially greater than that of the white; for instance, New York, in 1890, had a death rate of 28.5 per thousand for the white and 37.5 for the colored population; Chicago had 21.0 for white and 23.3 for the colored, and Charleston, S. C., had 24.7 for white and 53.9 for colored.

Of all deaths, about one-fourth are of infants under the age of one year, and somewhat more than one-third are of children less than five years old. The median age of the people of the United States is about 22 years; that is, one-half of the population is younger and the other half older than 22 years. It is therefore clear that the work of sanitary scientists and of boards of health should be especially directed to the improvement of the conditions that surround the young, in order that the median age of the population may be raised. If physicians were paid to watch over children when well, no doubt better results would be secured than under the present plan of relying upon cure by drugs rather than upon prevention by hygiene. Nowhere does the old adage, "an ounce of prevention is worth a pound of cure," better apply than in the case of sickness among the young.

The following table shows the number of deaths reported by the census enumerators in 1890 for the principal zymotic diseases and the corresponding percentages of the total reported number of deaths. As remarked above, these numbers of deaths are probably about 30 per cent too low, but the percentages in the third column may be regarded as fairly

DEATHS IN UNITED STATES IN 1890.

Cause of Death.	Number of Deaths.	Percentage.	Deaths per 1000 Living.
From all causes	875 521	100.00	18 00
Consumption	102 199	11.67	2.12
Pneumonia	76 496	8.74	1.57
Diarrheal diseases	74 711	8.53	1.54
Diphtheria	27 815	3.18	0.57
Typhoid fever	27 058	3.09	0.56
Malarial fever	18 594	2.12	0.38
Measles	9 256	1.05	0.19
Whooping-cough	8 432	0.96	0.17
Scarlet fever	5 969	0.68	0.12
Smallpox	398	0.04	0.01
The above 10 diseases	350 928	40.06	7.21

correct. Of course, these percentages vary greatly in different parts of the country. In the northern states diphtheria and typhoid fever are more prevalent than in the southern states, while the reverse is the case for malarial fever; thus in Vermont diphtheria caused 4.3 per cent of all deaths, while in Alabama it caused only 0.4 per cent. Diphtheria appears to be more prevalent in the cities, and typhoid fever in the rural districts. Consumption is more prevalent in the city than in the country, the highest rate being in Boston, Mass., where 16.5 per cent of all deaths were due to this disease in 1890.

The results just given show that about 40 per cent of all deaths in the United States result from causes of a zymotic or infectious character. Sanitary science seeks to decrease this percentage and ultimately to render these diseases as infrequent as smallpox now is. By so doing an annual death rate of 20 per thousand will be lowered to 12 per thousand, and the consequence will be a marked increase in the average age of the population. Death cannot be avoided, but it is the duty of man to prolong his life to the highest possible limit and to render it free from preventable disease.

What hygiene and sanitary science have done in the last two hundred years may be appreciated from the statement made by Farr, about 1870, that the annual death rate of the population of London in the latter half of the seventeenth century was nearly 80 per thousand, in the eighteenth century about 50 per thousand, and soon after the middle of the nineteenth century about 24 per thousand. To this may be added that for the decade 1871–1880 the average yearly death rate of London was 22.7, and for the decade 1881–1890 it was 20.5 per thousand inhabitants.

5. BACTERIOLOGY.

The germs which are the cause of zymotic disease are called Bacteria. These are of a fungoid nature, and belong to the lowest class in the vegetable kingdom. They are without color, very minute in size, and each individual consists of a single cell of organic matter within which watery fluid is contained. An individual bacterium is usually about one one-thousandth of a millimeter in diameter or thickness, and hence can be studied only under high microscopic power. The cells are spherical or cylindrical, the latter form being the more common; the word bacterium, meaning a little rod, is due to this shape. Propagation generally occurs by a contraction near the middle of a cell which then divides into two independent cells. When the conditions of temperature and food are favorable propagation goes on with great rapidity, and a few hours may be sufficient for one or two individuals to multiply into millions.

Bacteria are divided into three families, and each family into several genera, the family division being according to form. Thus the Micrococci are generally spherical and a group of these somewhat resembles a bunch of grapes; the Bacilli are the straight cylindrical rod-like cells; and the Spirilla are of curved or spiral form. The genera and species are distinguished partly by form and the method of propagation, but more definitely by the functions that they perform in the economy of nature or by the diseases with which they are connected.

The functions of bacteria are always connected with changes in organic matter, and these changes may be useful or injurious. The useful changes are those of fermentation and the alteration of decaying organic matter into harmless constituents. The injurious changes are those of disease by which healthy living organisms or tissues are brought into a state of disorder or specific poisoning. Hence bacteria are divided into two kinds according to their functions, the first being the useful or necessary class and the second the injurious or parasitic class.

The useful bacteria exist in the soil, the air, the water, and in the secretions and digestive systems of all animals and men. The change of milk into butter, or of wine into vinegar, is accomplished by their action. All fermentation results from chemical changes which are due to the development and life of these bacteria. All decay of organic matter and its change into harmless constituents is due to them. Without them the digestion and assimilation of food could not occur, and the highly organized life of man could not be possible. This class of bacteria, hence, is the most useful and important form of life, even though it be the lowest.

The parasitic bacteria, likewise, are found everywhere, endeavoring to increase and multiply by preying upon living organic matter. When the conditions become favorable their multiplication goes on with great rapidity, an injurious fermentation or poisoning results, the living tissues begin to decay, and disease results. Each species of parasitic bacteria produces its own specific disease, some attacking plants and trees, others animals and men, while still others prey upon man alone. Thus Bacillus amylovorus produces the apple blight, Bacillus anthracis causes anthrax fever in cattle, Bacillus tuberculosis causes consumption in men and animals, Bacillus diphtheriæ is always found in cases of diphtheria, Diplococcus lanceolatus in pneumonia, Bacillus coma in cholera, and Streptococcus variolæ in smallpox. The genus Bacillus ranks above all others in parasitic work, and over two hundred different species have been recognized and described.

Exactly how the operations of bacteria are performed and why such different diseases are caused has not been ascertained. Most bacteria simply float in the surrounding fluid, but others are capable of slight motion by means of hair-like appendages. The thin, colorless cells are capable of some pulsation, and it is by the secretion of fluid through the walls of these cells that their useful or injurious work is done. If by any means this process is interrupted, as has been done in some experiments by a dose of chloroform administered to a colony of bacteria, the processes of reproduction and secretion are stopped and the useful or injurious changes in the organic matter are completely suspended. It is known that the useful bacteria require a certain degree of moisture and heat, and that the presence of oxygen is necessary for their work. It is also known that sunlight prevents their development and that they are entirely destroyed by exposure for half an hour to a temperature equal to that of boiling water. Why it is that certain species cause disease and that others do not, or how it is that an almost infinitesimal amount of dangerous bacterial material may produce an epidemic, is not at all understood. But the study of bacteriology is not yet fifty years old; in half a century more our knowledge concerning it will doubtless be vastly more accurate both in regard to general principles and specific details.

6. Organic Matter.

Organic matter may be either vegetable or animal, living or dead. What constitutes the essence of life in the plant or animal is a deep mystery, but of the phenomena of life much has been learned. Living organic matter is that which is undergoing systematic change by the absorption of dead organic matter as food; at first each individual has a growth, and after this ceases disease or decay begins; in due time death ensues, and the dead matter is resolved into other forms and is absorbed by other living individuals. From the bacteriological point of view it is seen that the useful bacteria promote length of life, while the parasitic ones tend to shorten it.

Organic matter consists mainly of Carbon (C), Hydrogen (H), Nitrogen (N), and Oxygen (O). For instance, starch is $C_6H_{10}O_5$ and cane-sugar is $C_{12}H_{22}O_{11}$; albumen of the blood is $C_{72}H_{112}N_{18}SO_{22}$; wheat flour contains about 65 per cent of starch, 13 per cent of albuminoids, 2 per cent of nitrogen, together with cellulose, water, and ash phosphates. The body of an animal is similar in composition to its food, for the body has grown by the assimilation of that food. Carbon supplies heat to the body and also forms fat, while nitrogenous foods furnish the material for the muscles. In all organic tissue, and especially in the bones of animals, there are also found certain compounds of the metals potassium (K).

sodium (Na), and calcium (Ca), such as sodium chloride (NaCl), or common salt, and potassium nitrate (KNO_3), or saltpeter.

A plant absorbs water (H_2O) and carbon dioxide (CO_2) from the atmosphere and soil, and also nitrates of the metallic elements from the soil. A nitrite, it will be remembered, is the result of the combination of nitrous acid (HNO_2) with a metallic alkali, while a nitrate is formed by the stronger nitric acid (HNO_3); thus potassium hydroxide (KOH), or potash, acted on by nitrous acid yields potassium nitrite (KNO_2), but when acted on by nitric acid it yields potassium nitrate (KNO_3). From this food the plant grows, and in due time it becomes either food for animals and men or suffers decay. In the animal the organic matter is worked over into new forms, and it likewise finally becomes either food or decaying substance. The dead and decaying matter is then resolved back into carbon dioxide, water, and nitrates in the manner now to be described.

The following diagram represents dead organic matter as attacked by the oxygen (O) of the atmosphere. Here M represents potassium, sodium, and the other metals which are found as compounds in the dead organic matter. If the tem-

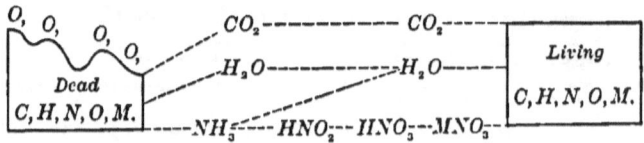

TRANSFORMATION OF DEAD INTO LIVING ORGANIC MATTER.

perature is sufficiently high the bacteria begin their useful work, and the first result is that the oxygen (O) combines with the carbon (C) to form carbon dioxide (CO_2). Next the oxygen proceeds to attack the hydrogen (H) and nitrogen (N), and ammonia (NH_3) is formed. By further attacks of the surrounding oxygen there are formed water (H_2O) and nitrous acid (HNO_2), which finally becomes nitric acid (HNO_3), and the action of this on the metallic compounds in the soil results in the nitrates (MNO_3). Then living organic matter picks up these constituents as the necessary food for its growth.

The most remarkable thing about this process of decay is that it requires the presence and active work of bacteria. In particular, the process of nitrification, or the oxidation of the ammonia to nitrous and nitric acid, is a chemical operation which is thought to be impossible without the operation of certain species of bacteria. At first their numbers are enormous, but as the nitrous acid changes to nitric they become less numerous, and when the nitrates have been fully formed few, if any, bacteria remain. Thus the work of these bacteria is done at the expense of their life; and it is thought that without this life and work no higher forms of life could be possible.

7. Filth and Disease.

In Art. 2 it was shown that the filthy habits and unclean surroundings of the people of the middle ages resulted in horrible diseases of an epidemic nature. The reason for this may now be understood. Disease is caused by bacteria, and wherever filth and decay abound there are bacteria in countless myriads. It is true that the bacteria which are transforming decaying matter into harmless constituents are doing good work, but it is also true that under conditions of abundant food the noxious as well as the useful bacteria find opportunity for growth and multiplication. Moreover, in decaying organic matter worms and insects abound upon which parasitic bacteria are preying. Thus filth causes the specific bacteria of germ diseases to increase and multiply, the surrounding atmosphere, water, and soil become impregnated with them, and then they make their attacks upon man. Further, the inhabitants of a community whose streets, houses, and persons are unclean usually do not have as nourishing food, systematic exercise, and refreshing sleep as those who live under good hygienic conditions. Consequently, their general power of resistance to disease is of a low order, and when an epidemic comes they readily fall before it. Each epidemic causes greater and greater debilitation, the diseases increase in virulence, and finally the horrible black death sweeps over the land. Forty-five epidemics of this plague occurred in Europe during the seventeenth century, and in the year 1665 it caused about 65 000 deaths in the city of London, which then had a population of only 200 000.

It is not at all necessary to accept the above reasoning in order to establish the connection between filth and disease, for this is done most effectively by facts. All statistics show that the rate of mortality is greater in the city than in the country, and that the highest mortality in a city is found in the tenement districts, where streets, houses, and people are unclean. Again, by the introduction of a pure water supply and the construction of a sewerage system it is found that the mortality from dangerous zymotic diseases is markedly decreased. Thus, at Manila, in the Philippine Islands, cholera was endemic twenty years ago, and frequently there were a hundred deaths in a single day, but since the introduction of a water-supply system it has almost disappeared. Again, at Danzig, in Germany, the annual death rate from typhoid fever during 1865–69 was 2.2 per thousand inhabitants; during 1871–75, after the introduction of water supply, it was 1.8; and during 1876–80, after the completion of a sewerage system, it fell to 0.4 per thousand. Similar examples in great number might easily be given to illustrate how the removal of filth causes disease to decrease.

The solid excrement of man is a dangerous form of filth, and particularly that from a person ill with fever. Three or four hundred years ago the usual method of its disposal in the cities of Europe was to throw it into the streets and trust to the rainfall to wash it away. Clothing infected with the discharges of sick persons was washed near public fountains or in streams from which drinking water was obtained, and as a consequence disease was spread in all directions. Such practices are now forbidden by law in Europe, but they continue in many countries. In India it is still customary to throw human excreta upon the surface of the ground and to drink water in which people have just bathed; as a consequence, cholera is always present and at times rages with great virulence. Yet there are localities in India where the people have adopted the European modes of living, and these are almost wholly free from cholera.

The experience and the statistics of the past two centuries teach most forcibly that the prevention of zymotic diseases in a town is to be effected in five principal ways: first, by proper hygiene of individuals as regards exercise, food, and cleanliness; second, by vaccination against smallpox and by disinfection and quarantine of persons having dangerous contagious diseases; third, by proper heating and ventilation of the houses; fourth, by the maintenance of a pure water supply; and fifth, by the removal of all filth from the town and its disposal in such manner that no contamination of the water supply may occur. The first and second of these methods are to be carried out by physicians and boards of health, the third is the province of the architect, the fourth and fifth belong to sanitary engineers. Binding all together are the regulations of law, which empower boards of health to suppress nuisances, punish those who pollute the water supply, and require the cooperation of all householders in disposing of their refuse and sewage.

8. Impure Air and Disease.

Many zymotic diseases are caused by infection communicated through the air. In some cases insects are the active agency; for instance, Texas fever in cattle has been shown to be communicated by wood-ticks, and probably the infection of malaria is propagated by the help of mosquitoes. Consumption is caused by dust, which conveys the specific bacteria of that disease from one person to another; the dried sputum of a consumptive patient has been shown to be an effective cause of such infection, and hence arose the recent prohibition by boards of health against expectorating on the floors of public conveyances and buildings. Measles, whooping-cough, and other children's diseases are undoubtedly communicated through the air of school-rooms, and perhaps even from house to house.

Pure air consists of 20.96 parts of oxygen, 79 per cent of nitrogen, and about 0.04 per cent of carbon dioxide. It is a mechanical mixture and not a chemical compound, and hence the percentages just given are liable to some variation. The proportion of oxygen may rise almost to 21 per cent on the sea or on high mountains, and it falls to 20.8 or 20.7 in crowded theatres and narrow alleys. The depressing effect of low oxygen and high carbon dioxide in an ill-ventilated lecture hall are known to all, and in such cases the animal matter of respiration, floating in the form of moisture or dust, is particularly liable to convey the germs of disease from the lungs of one person to those of many others.

The dust that floats in the air of a city street is composed of fine grains of sand, carbon, or smoke, animal manure of the streets, decaying vegetable matter, and many kinds of fungoid or bacterial growths. When it is considered that all these impurities are brought into contact with the blood in the lungs it is not strange that the infectious diseases are so readily spread by means of the air. It is a fact, however, that the liability to such infection is far less in the street than in the house, and this is doubtless due to the circumstance that the street air is in continual motion, and hence the dust is continually supplied with fresh oxygen to hasten the process of decay and thus reduce the number of bacteria and decrease their virulence. Within doors there is much less motion of the air, fresh supplies of oxygen are not furnished to the decaying dust, carbon dioxide and other products of respiration and transpiration are more abundant, and consequently the bacteria become more numerous and virulent.

It is hence important that the air of the streets should be pure, but still more important that the air within the houses should be kept so. To render the street air pure the street pavements should be maintained in proper cleanliness by frequent sweeping or washing, and by the removal of all garbage and decaying matter from around the houses. The width of streets should be such that the air may properly circulate and abundant sunlight enter. The stagnant air of narrow alleys and damp courts, where crowded tenements with damp cellars are found, is a most efficient propagator of disease, and in these localities the highest rate of mortality is found.

When the air of the streets is reasonably pure that of the houses will be so also, if the weather permits the windows to be open, and if the houses and their inhabitants be cleanly. But when artificial heat is required in the houses the problem of securing efficient ventilation is not so easy. The subjects of heating and ventilation must indeed be treated together in order to satisfactorily solve the problem. This is the province of the architect rather than of the engineer, and numerous excellent treatises discuss the question in great detail. Although the ventilation of houses is one of the branches of sanitary science, it is not strictly one of the departments of sanitary engineering, for the engineer's work is the execution of improvements for the general public rather than for the private individual. The architect erects the house and provides for its heating and ventilation, the engineer keeps the streets clean, builds the water-works, and provides for the removal of sewage. By the work of both professions the air within and without the house is to be kept as nearly pure as possible, and thus both the health of the family and of the public is to be conserved.

9. Drinking Water and Disease.

The water that we drink is assimilated into the blood in the same manner as food, and if it contain the bacteria of zymotic disease infection may be caused. After a period of incubation an inflammation of the intestines generally occurs, followed by fever and the other symptoms of the disease. Persons with weak constitutions who live in unclean surroundings are most liable to be attacked, while the strong and those who observe hygienic laws have the best chances of escape.

Cholera and typhoid fever are generally propagated through drinking water which has been infected with the characteristic bacteria of these diseases, although they may also be communicated by contagion with the clothing or person of a sick patient. Diarrheal and other intestinal complaints likewise result from the use of impure water. Two particular instances will now be given to illustrate the effects that follow the infection of a water supply.

In 1885 a severe epidemic of typhoid fever afflicted the borough of Plymouth, Pa., whose population was then about 7800. It first appeared on April 9, and there were 713 cases during April, 261 in May, and later 130 cases, making in all 1104 cases of sickness. The number of deaths was 114 or about $14\frac{1}{2}$ for each thousand inhabitants. An investigation clearly showed the infection to have been caused through the water supply in the manner that will be explained by the aid of the accompanying plan. A is the reservoir from which the town was generally supplied with brook water, and B, C, D are collecting reservoirs. In dry seasons this supply was insufficient, and in very cold weather it also failed through freezing; hence in such cases a pump P furnished river water instead. From March 20 to March 26 this pump supplied river water to the town, and then the pipes at *A* were thawed out, the pump was stopped, and the brook water admitted to the mains. At *H* is a house where from January to March a patient lay ill with typhoid fever contracted probably in December at Philadelphia. The excreta of this patient were thrown out upon the snow near the edge of the brook. On

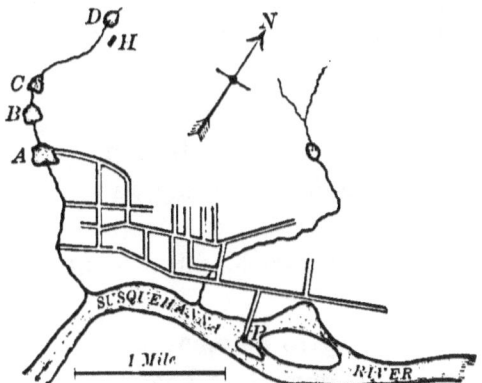

PLAN OF WATER SUPPLY OF PLYMOUTH, PA.

March 24 a thaw began and continued rapidly for a week or more, so that the infection doubtless reached the town approximately at the beginning of April. As the period of incubation of typhoid is twelve or fourteen days, this brings us to April 10-15, when cases were rapidly breaking out in Plymouth. Thus one case of typhoid multiplied into 1104, produced 114 deaths, and caused a heavy expense to the community, the amount distributed by a relief committee being $23 723, while the total loss has been estimated at $115 000.

In 1892 a terrible cholera epidemic visited the city of Hamburg, in Germany. Adjacent to Hamburg, and forming with it one continuous city, are the distinct municipalities of Altona and Wandsbeck. Hamburg used the unfiltered water of the river Elbe, Altona used the same water but thoroughly filtered it through sand beds before delivery into the pipes, while Wandsbeck derived its supply from a lake. On August 16 cholera broke out and 7427 cases occurred in August, 9341 in September, and 181 in October. The total number of deaths was 8976, the deaths in Hamburg being 134 per thousand inhabitants, in Altona 23, and in Wandsbeck 22. In Hamburg the disease prevailed in epidemic form, but the boundaries between it and Altona and Wandsbeck formed the lines beyond which the epidemic as such did not extend. In one street which for a long distance formed part of the boundary there was cholera on the Hamburg side, whereas the other side was free from it. Investigation showed that the cholera bacillus was found in the water of the Elbe, and there can be no doubt but that the people of Hamburg were infected by this water, while the filtration of the Altona supply rendered it harmless.

It would be very easy to fill many volumes with instances showing how epidemics of typhoid fever and cholera have been produced by impure water, and how the introduction of pure water has diminished the mortality from these diseases. The reports of sanitary commissions and boards of health abound in such illustrations, and the statement made at the beginning of this article is as thoroughly established as the fact that smallpox is prevented by vaccination. The water used by a town, therefore, must be maintained pure in quality in order to prevent typhoid and other allied zymotic diseases, and from time to time it should be examined by chemical and biological methods, in order to ascertain whether changes are occurring that may prove threatening or injurious.

10. Matter in Natural Waters.

Water as found in nature always contains some inorganic and organic matter. Both of these occur in two forms, in suspension or in solution. Suspended matter is that which is floating in the water, like inorganic sand or silt; dead organic matter, like leaves, sticks, feathers, and animal tissue; or live organic matter, like desmids, algæ, and insects. Suspended matter may be in large part removed from water by allowing it to settle in reservoirs or by passing it through filtering screens. Dissolved matter, on the other hand, is that which is so thoroughly in solution that it cannot be removed by settling or by screening, and the nature and amount of this gives to different drinking waters different distinctive characteristics. In a river the suspended matter during periods of high water may be ten or twenty times as great as at low water, while the dissolved matter is rarely twice as great; in spring water the variations are quite small.

It is clear that much suspended matter, whether inorganic or organic, renders water undesirable and perhaps unfit for domestic use. Silt in suspension acts unfavorably upon the intestines, while organic matter in suspension is generally in the process of decay and hence may cause zymotic disease. A water which is turbid in appearance generally contains much suspended matter and this should be removed by settling or filtration before it is suited for a public supply. The methods for doing this will be described in the next chapter.

All natural waters contain in solution certain gases, such as oxygen, nitrogen, and carbon dioxide; these are not at all injurious to health, but are often beneficial. The presence of little oxygen is generally an indication of impurity, for if the water contains decaying organic matter the dissolved oxygen is used for its combustion and nitrification. The agitation of water, in order to introduce oxygen into it, is in fact one of the methods of purification that will be described later. Dissolved nitrogen is of little influence. Dissolved carbon dioxide is usually beneficial in increasing palatability, but if the water flows through a limestone country this gas causes the formation of carbonates and an increase in the hardness of the water.

A water is said to be "hard" when it contains in solution the carbonates and sulphates of calcium or magnesium. The effect of these is to improve the water in taste, but when it is used for washing these carbonates and sulphates must be decomposed by the action of soap before a lather can be formed. Hard waters, hence, are more expensive in domestic use than soft ones. Hardness is said to be "temporary" when it can be removed by boiling, and this is the case when carbonates alone are present; it is said to be "permanent" when boiling is not sufficient, but soap or other chemical means are required, and this is the case when sulphates are present. Many waters contain both carbonates and sulphates, and for these the term "total hardness" expresses their combined influence.

Sodium chloride, or common salt, is found in all natural waters, the proportion being much higher near the sea coast or in the vicinity of salt beds than in inland surface and ground waters. This is not at all injurious to health, but if the proportion is found to increase in a well or water supply it is an indication of sewage contamination, for sewage contains some of the salt which is eaten by men and animals. In chemical analyses chlorine is determined instead of the sodium chloride, the amount of the former being always proportional to that of the latter.

Ammonia in water indicates the presence of organic matter in an advanced stage of decay, as shown by the diagram in Art. 6. The amount of this, though very small and not injurious to health, is a valuable indication of what is going on in the water. It is determined in two forms, called "free ammonia" and "albuminoid ammonia." Free ammonia is that which has been actually set free in the water in the process of decay, while albuminoid ammonia is that which has not been set free but is liable to bcome so under further active attacks of oxygen. The sum of the two gives an indication of the total amount of organic matter in the water, but free ammonia usually indicates greater danger than does albuminoid ammonia.

Nitrates are the final result of the nitrification action, and the amount of these gives much valuable information. Nitrites are an incomplete result of the same process. The examination of a water usually determines "nitrogen as nitrites" and "nitrogen as nitrates," and the larger the amount of these the greater the amount of organic matter previously in the water. If the ammonia be very low and the nitrates high the water has been completely purified; if the reverse is the case the decaying process is going on and the water is dangerous. Nitrites indicate danger, for complete purification is not effected until the nitrification has resulted in the formation of nitrates.

The useful bacteria employed in the process of decay are always accompanied by parasitic companions which are eager to prey upon living beings. These noxious bacteria are the dangerous things in potable water. A bacteriological analysis of water is hence most desirable: if few or no bacteria are found the water may be regarded as one which cannot cause zymotic disease; if many are found, it is liable to do so. In all cases the biological analysis is of great value in enabling the chemical one to be rightly interpreted.

11. Chemical Analysis of Water.

It is not here intended to give a description of methods of chemical analysis which will enable a student to perform them, but merely an account which will furnish him with such intelligent ideas that he can better understand and interpret the analyses. Of course, the first thing to do is to get the water, and about a gallon is needed. This should be collected in a glass vessel with a glass stopper, and it is essential that the jar should have been sterilized in order to remove all organic matter and destroy all bacteria that may have been within it.

Water should be analyzed soon after having been collected, for if it be impure the processes of oxidation and nitrification may to a certain extent render it purer after a few days. It should not be strained or filtered, as the object of the analysis is to determine the purity of the sample collected.

As the quantities to be obtained by analysis are very small, they are expressed in parts per million, and always by weight. One part per million is hence the same as one milligram in a kilogram; thus, if chlorine be given as 6.4 parts per million, this means that one kilogram of water contains 6.4 milligrams of chlorine. Some chemists unfortunately express the results in parts per 100 000 instead of in parts per million.

Hardness.—This is expressed by the number of parts of calcium carbonate ($CaCO_3$) contained in one million parts of the water; or if calcium sulphate ($CaSO_4$) is present it is reduced to an equivalent amount of calcium carbonate. To determine this, one milligram of calcium carbonate is dissolved in one liter of distilled water, and the amount of a standard soap solution which will form a permanent lather with this water is determined. Then the standard soap solution is applied to the water under analysis, and the amount of it needed to form the same lather is proportional to the parts per million of calcium carbonate which this water contains. Rain water has a hardness of about 5, river waters from 50 to 100, while limestone waters have 200 or more parts per million.

Total Solids.—This term indicates the total solid matter, both organic and inorganic, in one million parts of the water. About one hundred grams of water are placed in a platinum dish of known weight and the whole accurately weighed. The water is then entirely evaporated by boiling, and the dish, after cooling to the original temperature, is again weighed; this last weight minus the original weight of the dish gives the total solids in the water used. If the dish be heated to redness the organic matter is burned out, while the remaining ash shows the inorganic matter. Rain water may have the total solids as low as 20, while ground waters may run up to 500 parts per million.

Chlorine.—This is determined by the fact that, if a solution of sodium chloride be colored yellow with potassium chromate and silver nitrate be added, white silver chloride will be precipitated until all the chlorine is used up, and then the red color of silver chromate is seen. The solution of silver nitrate is first standardized, and then, being applied to the water under analysis, the amount required to produce red color is proportional to the parts per million of chlorine contained in that water. Inland brook and spring waters have from 1 to 5 parts per million of chlorine, but near the sea coast the proportion may be several times as great.

Free Ammonia.—This is also determined by a color test, a certain solution of mercury chloride, called Nessler's solution, yielding a brownish yellow coloration with the smallest trace of ammonia. The color produced by the water under analysis is compared with those produced by standard ammonia solutions, and thus the parts per million of free ammonia in the given water are known. Water having 0.05 parts per million is probably very pure; if above 0.1 it is suspicious, and perhaps dangerous.

Albuminoid Ammonia.—This is determined by first distilling off all the free ammonia and then adding an alkaline solution of potassium permanganate to oxidize the nitrogenous organic matter remaining in the water. By this oxidation ammonia is set free, and this is measured by the Nessler solution in the same manner as before. This ammonia is called albuminoid because albumen gives off ammonia when treated with potassium permanganate. The amount of albuminoid ammonia in good water is also very small, 0.05 being a very low figure and 0.50 a high one.

Nitrogen as Nitrates and Nitrites.—The amounts of these are also ascertained by color tests, which will not here be described, comparison being made by standard solutions. In order to give full information the amount of nitrogen as nitrates and nitrogen as nitrites should be reported separately, the former showing a more perfect nitrification than the latter. In good waters the amount of nitrogen as nitrates may be as high as 1 or 2 parts per million while that of nitrogen as nitrites is a mere trace.

Oxygen Consumed.—This term means the amount of oxygen absorbed by the water from potassium permanganate, which is added gradually until the purple color remains permanent for ten minutes. The oxygen set free from the potassium permanganate is absorbed in oxidizing the organic matter, and thus the greater the amount consumed the more impure is the water. Less than one part per million indicates purity, while as high as 4 or 5 probably indicates danger.

The determination of hardness is needed mainly in the examination of a proposed water supply. For an established supply the determinations above noted are those generally made in order to judge of the degree of purity, and the cost of such an analysis should be less than $20. Total solids are often reported in two parts; the one indicating organic matter is sometimes called "loss on ignition," while the one showing the inorganic matter is termed "fixed residue." Many analyses omit the determination of oxygen consumed.

12. Biological Analysis of Water.

The general properties of a water as regards color, odor, and taste are usually reported both in chemical and biological work, and such reports are of value as indicating suspicion only. The words brown or yellow, as applied to color, and vegetable, fishy, and mould, as applied to color and taste, certainly give unpleasant impressions. It goes without saying that water having a turbid appearance and unpleasant smell is suspicious, but it does not necessarily follow that it is dangerous. A statement that water is free from color and odor has, however, no especial significance, for some noted epidemics of cholera and typhoid fever have been caused by clear and sparkling waters.

A microscopic examination is valuable in ascertaining the kinds of suspended organic matter which the water contains. The dead organic matter may be found to be either vegetable or animal, and it may be also ascertained what particular plants or animals furnish the refuse. The living organic matter will in general be found to be certain species of the fresh-water algæ of the desmid and diatom families.

Desmids are of greenish color, while the diatoms are brownish and have a somewhat silicious structure. Desmids have not given trouble in water supplies, but the diatoms often do so, as when they decay a fishy or pig-pen odor is produced. The genera Crenothrix and Volvox, in particular, often cause these disagreeable odors, but it cannot be definitely said that they are the cause of disease. These families are the lowest ones in the vegetable kingdom which have color, the still lower class of bacteria being colorless. They can easily be recognized by a microscope magnifying about 200 times, while the bacteria require microscopes of the highest power. In complete biological analyses records are made of the number of each genus of diatoms and other algæ contained in one cubic centimeter of the water.

The bacteriological examination is, however, the important part of a biological analysis. The object of this is to determine the number of bacteria present in one cubic centimeter of the water, and from this number to judge of its purity. If no bacteria are found it is inferred that the water cannot communicate zymotic disease, and the larger the number found the greater is its liability to do so. Good water contains less than 100 bacteria per cubic centimeter.

In order to count the bacteria, a culture jelly, consisting of gelatine, albumen, and extract of beef, is prepared in order to furnish food upon which the bacteria may feed and multiply. One cubic centimeter of water is then thoroughly mixed with about 10 cubic centimeters of liquefied culture jelly, and the whole spread out in a thin layer upon a sterilized dish to harden. Each individual bacterium then begins to eat, to divide by fission, and the multiplication continues until a colony is produced which is visible to the eye. After about 48 hours the number of colonies is counted, and thus the number of bacteria in one cubic centimeter of the water is known. When the number is large a plate of glass ruled into squares is used, and the count is made over a certain fractional part of the dish.

By further examination with the microscope the different species of bacteria may be ascertained, but this is rarely done, as it requires expert skill of a high order. In fact, the entire bacteriological analysis requires much skill, in order to prevent the introduction of bacteria upon the gelatine from other sources than that of the water under analysis.

Another bacteriological method is to inoculate a rabbit with bacteria from the culture jelly and then to observe whether the animal manifests any symptoms of zymotic disease. This is usually ascertained by killing the animal after the expiration of the period of incubation and examining its intestines to ascertain if the characteristic inflammation of typhoid fever be present. This, of course, requires special skill and cannot generally be carried out except in biological laboratories.

A popular test for water, known as Heisch's sugar test, and which may easily be made by any one, will here be noted, as it depends entirely upon the development of bacteria. Let a pint bottle of colorless glass, and having a glass stopper, be thoroughly cleaned by immersion in boiling water for half an hour. Let it be entirely filled with the water to be examined, and let a teaspoonful of white sugar be added.

Then let it be exposed to the light in the window of a warm room for a week or ten days. If the water becomes turbid it is open to grave suspicion, but if it remains clear it is almost certainly safe.

13. INTERPRETATION OF ANALYSES.

The chemical methods of analysis have been longer known and are better systematized than the biological ones; they are found recorded and interpreted in the annals of sanitary science, and hence on the whole give more important and extended information than the biological analyses. Undoubtedly both methods of analysis will go hand in hand in the future, the biological work will be further perfected, and each method will be found necessary to supplement the other.

It must not be supposed that a single chemical analysis can give decisive information as to whether a water is good or dangerous. In certain cases it may do so, but in most cases the interpretation of the results cannot be made unless the source from which the water was taken is known. This is due to the circumstance that the substances determined in the chemical work are not poisonous, but that they are merely indicative of the amount of organic matter in the water. Organic matter does not cause disease unless in such a state of decay that bacteria are at work, and whether or not this is the case depends upon the physical surroundings. A river water high in albuminoid ammonia and low in nitrates may sometimes be less injurious than a spring water where the former is low and the latter high. Chlorine gives little information unless the normal chlorine of the surrounding region is known. In short, a knowledge of the topography and sanitary conditions of the source of supply are absolutely essential in order to interpret satisfactorily a chemical analysis. After reading the next chapter, in which the qualities of different kinds of waters are discussed, the truth of this statement will be better appreciated.

There is a widespread idea among the public that a chemist should not know the origin of a water, in order that he may be unbiased in giving an opinion. Probably this arises from the impression that a chemist judges of the influence of the ammonias and nitrates in the same manner that he does regarding such poisons as arsenic and lead. From what has been said in this and the preceding articles, it is seen that this is not at all the case. These substances are not poisons, but merely indications of the amount of organic matter in the water, and no reliable chemist will venture to give an opinion as to the purity of water without knowing its source and surroundings.

In the case of water infected with the bacteria of cholera or typhoid fever chemical analysis will generally fail to give any indication of such infection. This is because the amount of infected matter and its accompanying bacteria is generally so very small as not to add appreciably to the amount of organic matter previously present. In an experiment by Latham the dejections of a cholera patient were added to pure water in sufficient amount to impart the disease to any one who should drink it, and yet chemical analysis entirely failed to discover any essential difference between the two waters. In such a case a single bacteriological analysis may be far more valuable than hundreds of chemical ones, and in general each succeeding year adds to the importance of such examinations.

In most books there are given what are called "standards of purity," that is, certain limits beyond which a water is to be regarded as dangerous in use, or certain figures which are regarded as reasonably safe. Such standards necessarily differ in different localities and with different kinds of water, and on the whole can be regarded only as expressions of individual opinion. For instance, the rules of the Michigan State Laboratory of Hygiene, in force in 1897, give the following parts per million as maximum allowable limits for drinking water: Hardness, 50; Total Solids, 500, of which the inorganic should not exceed 200; Chlorine, 12.1; Free Ammonia, 0.05; Albuminoid Ammonia, 0.15; Nitrogen as Nitrates, 0.9; Nitrogen as Nitrites, a trace; Oxygen consumed, 2.2; Bacteria, "no toxicogenic germs, as demonstrated by tests on animals." These standards are severe ones, and it may be said again that the true method of judging the quality of a water is not by observing whether one of the determinations reaches or surpasses a certain limit, but by studying the analysis as a whole in the light of the source and surroundings of the water. It is not to be expected that the student can do this at the outset, for much experience and judgment are needed in order to announce confident conclusions. Indeed, these should not be generally announced without a study of several analyses of the water taken at different times, for it is found that the proportions of total solids, ammonias, and nitrates are different at different seasons of the year.

The following figures will assist the student to form an idea of the difference between the analyses of a good water and a very impure water, like sewage:

		Good Water.	Sewage.
Total solids,	parts per million	50.0	700.0
Organic matter,	" "	30.0	200.0
Inorganic matter,	" "	20.0	500.0
Chlorine,	" "	3.0	40.0
Free ammonia,	" "	0.010	25.000
Albuminoid ammonia,	" "	0.100	10.000
Nitrogen as nitrates,	" "	0.200	0.100
Nitrogen as nitrites,	" "	0.000	0.005
Oxygen consumed,	" "	0.5	40.0
Bacteria, per cubic centimeter		25	1 000 000

These may be called typical analyses, because they are not actual ones, but express rough average figures which are intended to show the marked differences between the two types of water. The decimal places are carried out in each case as far as usual in analyses, but as these have here no significance they are filled with ciphers. Of course, different kinds of good water and of sewage will furnish analyses that may differ very much from these typical ones.

14. Results of Sanitary Science.

In the preceding pages have been briefly outlined those elements of sanitary science which are essential to the study of sanitary engineering. The historical notes show what was done in early times and point out how the filthy habits of the people of the middle ages led to direful epidemics of plague. The classification of diseases, the statistics of mortality, and the section on bacteriology explain the modern theory of the transmission of zymotic disease by means of organic germs. The changes in organic matter during decay are next discussed, and it is seen that the chemical operations require the presence of bacteria to secure successful purification. Thus the methods by which filth originates disease and causes its communication to men through air and water are rationally understood. It then follows that streets and houses should be well ventilated, that a pure supply of water should be maintained, and that effective drainage and sewerage should remove all filth from the town. The chemical and biological methods of analyzing water are then taken up, and thus the foundation is laid for the discussions of the qualities of water and the methods of its purification which are to be given in the next chapter.

Many instances have been given showing how zymotic diseases have been lessened in extent and intensity by the observance of the principles of sanitary science. There still remains, however, the question as to what extent the average age of the community has been increased by the great reforms of the nineteenth century. In order to answer this question a discussion has been made by the author of the figures given in five censuses of the United States, and the conclusions derived will now be presented. Notwithstanding the many errors and imperfections of a public census, owing to incorrect answers given by the people and to carelessness of the enumerators, these errors are governed by definite laws, which are the same in all the censuses. Accordingly, the enumeration of the people by ages may be confidently used to furnish results from which accurate comparisons and conclusions can be drawn.

The following table gives the median age of the people of the United States at each of the five decennial censuses. The median age is an age such that one-half of the population is less than it and the other half greater than it. Thus in 1850 one-half of the total population was under and one-half over 18.28 years. The second column of the table shows that there was a gain of 3.1 years in median age of all classes of population in the 40 years from 1850 to 1890, the third shows that the gain in median age of the white population was slightly greater, and the fourth shows that the increase for

MEDIAN AGES FOR THE UNITED STATES.

Year of Census.	All Classes.	Whites.	Colored.	Native Whites.
1850	18.28	18.60	16.48	
1860	18.87	19.31	16.53	
1870	19.65	19.90	17.67	16.21
1880	20.45	20.93	17.47	17.80
1890	21.38	21.95	17.65	18.87
1900	(22.3)	(23.0)	......	(19.9)

the colored population was only about 1.2 years. Fair conclusions cannot be drawn from the second and third columns, on account of the influence of immigration, and accordingly the figures for the native whites have been added in the last column as far as available. These show a higher increase still, and the general conclusion follows that the median age of the white population has now a rate of increase of almost one year in each decade; therefore, for 1900 the predictions may be made which are given in parentheses in the last line. This increase in age is the triumphant result of the progress of sanitary science during the latter half of the nineteenth century.

15. Exercises and Problems.

The following series of exercises and problems is presented for the use of students in engineering colleges, and they will prove of great value in giving habits of thought and investigation to all who carefully perform them. Some of them may be solved by reference to dictionaries and cyclopedias, but others will require the consultations of the special articles that are mentioned. The numerical problems of this chapter involve only an elementary knowledge of arithmetic, chemistry, and physics. The number prefixed is that of the article of the text which is especially related to the exercise or problem.

1. What are the meanings of the words Etiology, Antitoxine, Demography, Pathology, Toxicology, Zymosis?

2 (a) What meats may be eaten and what may not be eaten according to the sanitary code in Leviticus, xi, and Deuteronomy, xiv?

2 (b) Describe the Roman sewer called Cloaca maxima, and the Roman aqueducts called Aqua virgo and Aqua Claudia.

2 (c) Read chapter vii of Volume II of Draper's Intellectual Development of Europe; describe the condition of English life in the latter part of the thirteenth century; also how syphilis spread over Europe.

3 (a) Describe smallpox, and state the methods of its prevention by inoculation and by vaccination.

3 (b) Is consumption an infectious disease? See Harper's

Magazine, March, 1894, or Prudden's Story of the Bacteria (New York, 1889).

4 (*a*) Consult the Compendium of the Eleventh Census, Vol. I, p. 468, and Vol. II, p. 20. Assuming the death rate for 1890 to be 18.0 per thousand for the total population, what was the death rate for the white population and for the colored population?

4 (*b*) How many deaths from consumption and smallpox occurred in the United States in 1870, 1880, and 1890?

5. Give sketches showing the characteristic forms of the three classes of bacteria. What is the name of the germ that causes typhoid fever?

6 (*a*) If a barrel of cane sugar ($C_{12}H_{22}O_{11}$) weighs 342 pounds, show that the number of pounds of carbon, hydrogen, and oxygen is about 144, 22, and 176.

6 (*b*) What results when nitric acid (HNO_3) acts upon soda ($NaOH$)?

7 (*a*) Consult Transactions of Seventh International Congress of Hygiene and Demography (London, 1892), Vol. XI, p. 136, and give an abstract of Dhurandhar's account of the sanitary condition of villages in the Bombay district of India.

7 (*b*) Give an abstract of an article by Jordan and Richards on the Nitrifying Organism in Part II of Experimental Investigations by the State Board of Health of Massachusetts (Boston, 1890).

8. In an unventilated heated room where is the warmest air and where is the air containing the most carbon dioxide? Where should fresh air be admitted and where should the foul air be taken out in order to give the most effective ventilation?

9. Consult the recent reports of one of the State Boards of Health, and give instances of disease communicated by bad water.

10. Consult Part I of Experimental Investigations by the State Board of Health of Massachusetts (Boston, 1890), and give an account of the work for determining the normal chlorine in that State.

11 (*a*) If a water has a temporary hardness of 83 and a permanent hardness of 42 parts per million, show that its total hardness is the same as that caused by one ounce of carbonate of lime dissolved in 8 cubic feet of water.

11 (*b*) In a platinum dish weighing 43.2675 grams 100 cubic centimeters of water are evaporated and it then weighs 43.3102

grams. Show that the total solids in the water are 427 parts per million.

12 (*a*) Consult Rafter's article on Purity of Water Supplies in Vol. XXI of Transactions American Society of Civil Engineers, and give instances of trouble caused by Crenothrix and Volvox.

12 (*b*) Make sugar tests of three samples of water one of which is known to be impure; at the close of the experiment observe also the odor from each sample.

13 (*a*) Consult Drown's article on Interpretation of Chemical Analyses of Water in the volume cited in question 10, and endeavor to interpret the analyses of normal and polluted waters given on page 541.

13 (*b*) Consult Fox's Sanitary Examinations of Water, Air, and Food (London, 1878), and give examples of mistakes made in furnishing opinions based on chemical analyses only.

14 (*a*) Consult the census reports of the United States and show that the percentage of the total population over $59\frac{1}{2}$ years was 4.13 in 1850, 4.29 in 1860, 5.02 in 1870, 5.64 in 1880, and 6.21 in 1890.

14 (*b*) Plot the figures in the first and second columns of the table of the median ages and show how the age 22.3 for 1900 is inferred from those preceding.

15. Define species, genus, family, class. What are the classes in the vegetable kingdom, and to which class do the bacteria belong? What discoveries were made by Pasteur? Who discovered the bacillus of cholera? What is the antitoxine method for the prevention or cure of zymotic disease?

Chapter II.

WATER AND ITS PURIFICATION.

16. Rainfall.

The water which has been evaporated from land and ocean is precipitated in the form of rain when the temperature of the atmosphere is lowered by cooling winds. As the rain falls it collects impurities from the dust of the atmosphere and from the surface of the land. Running over the surface, it forms swamps, brooks, and rivers; percolating into the earth, it appears again as springs and wells. Thus all water supply is primitively due to rainfall.

Rainfall is measured in inches or centimeters of vertical depth. A rain gage consists of an open vessel for collecting the falling water and a cylindrical glass tube of smaller cross-section in which the heights may be more easily read. If the area of the cross-section of the tube be one-tenth of that of the vessel one vertical inch of water in the vessel occupies ten inches in height in the tube, and 0.01 inch in the vessel is 0.10 inch in the tube; the graduation of the tube is made so as to directly give the actual amount of rainfall. The simplest way is to pour the water from the vessel into the tube by the help of a funnel, but in the best work both vessel and tube should form one instrument. For precise observations self-registering gages are used, so that the intensity of rainfall at each instant may be known. Snow and hail are melted and the resulting water included in the record as rainfall.

The frigid zone has the least rainfall and the torrid zone the greatest. At the equator the average annual rainfall is about 100 inches, at latitude 40° it is about 40 inches, and at latitude 60° about 20 inches. There are, however, some regions in the temperate zones where practically no rain ever falls, as in middle Egypt, and others where the annual rainfall is 500 inches, as in the Cossyah Mountains of India. The rainfall in any locality depends upon the character of the winds and upon the neighboring mountains and oceans.

In the United States there is an extensive region, formerly called the Great American Desert, where the mean annual rainfall does not exceed 15 inches; this embraces the states of Arizona, Nevada, New Mexico, Colorado, Utah, Wyoming, and Montana, with parts of adjacent states. The least annual rainfall for any state is $7\frac{1}{2}$ inches in Nevada. In all this region irrigation is necessary for the pursuits of agriculture, the water falling in the wet months being impounded for use in the dry season.

The states having the heaviest annual rainfall are those on the Gulf of Mexico, the mean amount being from 50 to 55 inches; Florida and Louisiana stand highest, with about 55 and 54 inches, respectively, but there are regions in these states where the annual rainfall exceeds 60 inches. The maximum annual rainfall in the United States is found, however, near Puget Sound, in Oregon and Washington, where the mean is 50 inches or more per year, although for the entire surface of these two states it is only about 40 inches.

At any place the rainfall in a given year is liable to vary considerably from the mean for several years. Thus, the mean annual rainfall at Philadelphia, Pa., for the ten years 1881–1890 was 39.6 inches, but the highest annual rainfall was 50.8 inches in 1890 and the lowest 33.4 inches in 1885, the variations of these from the mean being 28 and 16 per cent, respectively. Similarly, at Denver, Col., the variations of the maximum and minimum annual rainfall from the mean during the same years were 40 and 46 per cent. Whenever the water supply of a town depends directly upon rainfall, as it does in most cases where collecting and storage reservoirs are used, the minimum annual rainfall is a factor of much greater importance than the mean.

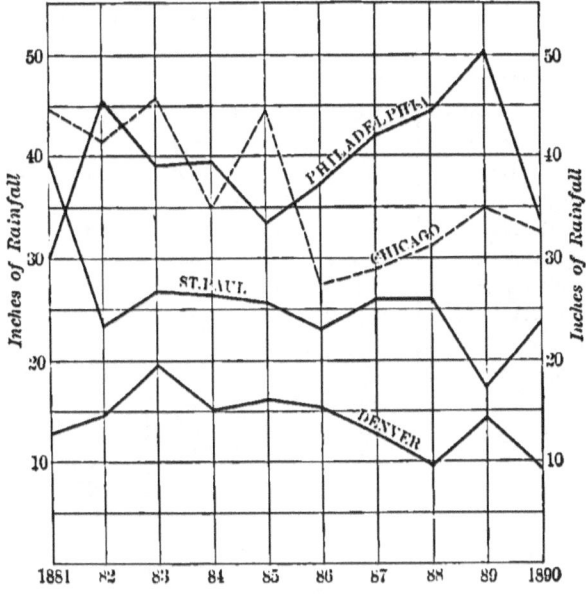

RAINFALL AT FOUR AMERICAN CITIES.

The distribution of the rainfall throughout the year is very different at different places. As a rough general rule the summer rainfall is the greatest and the autumn rainfall the least, but this is reversed in a few states. The following table, abstracted from a larger one compiled by the United States Weather Bureau, gives a general idea of the mean seasonal variation in different parts of the country. In Massachusetts the rainfall is equally distributed throughout the year, but in California the winter rainfall is 10 times as great as that in the summer. The variation in rainfall by seasons and months must be carefully regarded in planning storage systems, and for this purpose the rainfall records of each special locality should be obtained and discussed; in these plans it should be the aim to secure an abundant supply during the driest months.

MEAN RAINFALL IN DIFFERENT STATES AND SEASONS.

States.	Spring Rainfall. Inches.	Summer Rainfall. Inches.	Autumn Rainfall. Inches.	Winter Rainfall. Inches.	Annual Rainfall. Inches.
Massachusetts,	11.6	11.4	11.9	11.7	46.6
New York,	8.5	10.4	9.7	7.9	36.5
Pennsylvania,	10.3	12.7	10.0	9.5	42.5
Virginia,	10.9	12.5	9.5	9.7	42.6
South Carolina,	9.8	16.2	9.7	9.7	45.4
Alabama,	14.9	13.8	10.0	14.9	53.6
Louisiana,	13.7	15.0	10.8	14.4	53.9
Kentucky,	12.4	12.5	9.7	11.8	46.4
Illinois,	10.2	11.2	9.0	7.7	38.1
Minnesota,	6.5	10.8	5.8	3.1	26.2
Nebraska,	8.9	10.9	4.9	2.2	26.9
Colorado,	4.2	5.5	2.8	2.3	14.8
Montana,	4.2	4.9	2.6	2.3	14.0
California,	6.2	0.3	3.5	11.9	21.9
United States,	9.2	10.3	8.3	8.6	36.3

In some discussions regarding reservoirs, and also in those of sewerage systems, the maximum daily and hourly rainfalls are matters of importance. At Boston, Mass., the greatest recorded rainfall in 24 hours prior to 1892 was 8.9 inches; at Philadelphia, Pa., 5.2 inches; at Charleston, S. C., 8.3 inches; at New Orleans, La., 8.9 inches; at Nashville, Tenn., 5.2 inches; at Chicago, Ill., 5.6 inches; at St. Paul, Minn., 3.7 inches; at Denver, Col., 6.5 inches; and at San Francisco, Cal., 4.7 inches. With regard to the maximum rainfall in one hour it may be said that 1 inch is a heavy precipitation, and that 2 inches is of rare occurrence at any given locality; on August 3, 1898, however, 3.8 inches fell in one hour at Philadelphia, Pa., and 0.9 inches of this fell in ten minutes.

17. Evaporation, Run-off, and Percolation.

After the rain has fallen a part of it runs off into the brooks and rivers and another part percolates into the soil. Evaporation immediately begins both from the land and water surface, and this continues until all the rainfall is ultimately evaporated into the atmosphere, where it is condensed into clouds and falls again as rain. For any particular watershed which supplies a reservoir, however, the evaporation is less than the rainfall, while the run-off may be impounded for the purposes of irrigation or water supply.

Experiments on evaporation are made by placing watertight pans at the level of the ground and noting daily the variations in depth, together with the rainfall. On the surface of a reservoir or lake similar experiments may be made by floating boxes. It is found that the evaporation from water surfaces is greater than that from the land, that it is greater in dry and desert regions than in cultivated ones, greater in low lands than on mountains, and that it increases with the temperature of the air and the velocity of the wind.

In the Atlantic states the annual evaporation from land surfaces may be regarded, on the average, as about 40 per cent, and that from water surfaces as about 60 per cent, of the annual rainfall; in low and level localities these percentages are much increased, while for high regions and steep slopes they are decreased. In the arid regions west of the Rocky Mountains the evaporation from water surfaces may be several times as great as the rainfall. In the first region nearly one-half of the annual rainfall may be utilized for a water supply, while in the second region the percentage that can be rendered available is much smaller.

The amounts of run-off and percolation depend upon the

topography of the country and the nature of the soil. In a level country the run-off is small, while on steep mountain slopes it may be as high as 80 per cent of the rainfall. For an undulating cultivated country the annual run-off usually lies between 40 and 70 per cent of the annual rainfall, while the remainder goes into evaporation and percolation. To determine the run-off of any watershed above a certain place on a stream the flow of that stream is to be measured by the methods explained in treatises on hydraulics, and if such observations be continued throughout a year of average rainfall a fair value of the mean annual run-off is obtained. Should such observations be impracticable, an estimate may be made based on percentages of rainfall and evaporation at localities having similar topography and climatic conditions. An example of such estimates will be given in Art. 33.

The percolation is partly absorbed by the roots of plants and trees growing on the surface, but a large portion of it sinks to a greater depth and forms what is called the ground water. This ground water is that which appears in springs and in artificial wells, and it has a steady flow through the earth towards the streams and the oceans. When the geologic conditions are favorable some of the percolation sinks much deeper and forms those subterranean reservoirs which are called artesian waters.

The phenomena of rainfall, run-off, and percolation give rise to three classes of drinking water. First, there is rain water which is caught as it falls; second, the surface water of swamps, brooks, rivers, and lakes; and third, the ground water of springs and wells. These waters are now to be described and compared, their sources of pollution indicated, and the methods for their purification discussed.

18. Rain Water.

Rain water is frequently collected and used for a family supply, particularly in country districts. The roof of the house forms the collection area, and from this the water is led either to a tank in the garret or to a cistern below the surface of the ground. As the roof is liable to become covered with dust, it is not well to allow the water to enter the tank or cistern during the first hour or two of a storm. The tank or cistern must be provided with a waste-pipe to prevent overflow, but under no circumstances should this be connected with a cesspool or sewer. When due care is used rain water collected in the country is very pure, although those unaccustomed to it object to its taste, and its softness renders it the best of all waters for washing purposes. When collected in the city it is liable to contain smoke, dust, and organic matter, with the accompanying bacteria, to a far greater extent than in the country.

Chemical analyses of rain water show that the total solids are lower than in other kinds of water, say in general from 20 to 40 parts per million. Free and albuminoid ammonia are always found, the former being often higher than 0.050 and the latter usually less. Chlorine may be as low as 1.0, but is usually higher near the coast, where salt is blown in by winds from the ocean. Nitrogen as nitrates is very low in amount, say 0.05 parts per million or less. On the whole, a chemical analysis of water from an uncontaminated cistern usually indicates purity in all directions, except that sometimes the free ammonia appears to run too high; this is not necessarily derived from decaying organic matter, but may have been washed out of the atmosphere by the falling rain.

The average householder cannot, however, be trusted to exercise due care in the collection of rain and in maintaining the tank or cistern in good order. A tank is liable to become infested with fungoid growths, and a cistern is liable to be contaminated by household refuse, surface drainage, or leakage from the soil. Birds or rats may sometimes get into the roof pipes, and the organic matter from their dead bodies be washed into the tank or cistern. Those who have been present at the annual cleaning of a cistern will recall the unpleasant odor that usually attends that operation, and such odor is an indication of the decay of organic matter. A common custom in some country districts is to throw pieces of charcoal into the cistern in order to alleviate these odors, but it cannot be said that this in any way removes the suspicion of impurity. Although for washing purposes rain water is most excellent on account of its lack of hardness and the consequent economy in soap, it must be concluded that it is not well to use it for drinking water after it has been stored in house tanks and cisterns except when great care and vigilance have been exercised to prevent contamination.

Snow collects matter from the atmosphere more readily than rain, and hence water melted from the first snow of a storm is more impure than common rain water; likewise, city snow is not as pure as that of the country. After snow has fallen it absorbs impurities from the soil or from the roofs of houses, so that the amount of organic matter and free ammonia may be doubled in a day or two. In general, water produced by melting such snow is not wholesome, and the same may be said of ice which is formed from snow slush.

19. Surface Waters.

Surface waters include those of swamps, brooks, rivers, and lakes, and these differ greatly in regard to their characteristics. Swamp water is liable to be heavily charged with vegetable matter, but the flow in brooks and rivers causes a continuous improvement in quality, and when a lake is reached the purest surface water is found. This improvement in quality is effected in two ways: first, by settling or sedimentation, which removes the suspended matter; and second, by aeration or contact with the air, whereby oxygen is supplied to decompose and destroy both the suspended and the dissolved organic matter.

Swamp water usually has a high proportion of vegetable matter in the total solids, and a high proportion of albuminoid ammonia is also found. In boggy and peaty regions this gives a brown color to the water, but it fortunately happens that the vegetable matter is in a permanent state which resists further decomposition, so that sometimes these waters are noted for their keeping qualities and are well adapted to being taken on long sea voyages. When used for a public water supply the aeration due to pumping and flow is apt to cause this organic matter to decompose, and hence filtration is necessary. Swamp water has been used at Long Branch, N. J., Norfolk, Va., and other cities without unpleasant results; its softness renders it convenient for washing, but brook or river water is always to be preferred for drinking purposes when it can be had.

Brook water consists of the run-off of the surface, of the drainage of swamps, and of percolation from meadow and springy land. It is generally thought that the color and taste of brook water give a reliable indication of quality, but a brook having its source in farm-yards or swampy pastures may furnish water which is very unwholesome, even though it be clear and sparkling. With steep slopes and rocky beds purification continually goes on, and if further sources of contamination be absent good potable water may perhaps be found a mile or two from the suspicious sources. Brook water is soft except when the flow comes from limestone springs, the organic matter is usually lower than in swamp water, while the nitrates are usually higher. It forms a reliable public supply for hundreds of towns, being purified by natural sedimentation in collecting reservoirs and sometimes by artificial filtration.

A river is formed by many brooks, and the quality of river water differs mainly from that of brook water in the higher proportion of inorganic matter which has been collected from the soil in the flow. A river having towns upon its banks is liable to become polluted by the drainage and sewage that run into it, and at present one of the important sanitary problems is how to dispose of the refuse of towns without polluting the streams. In Europe this subject has received much attention, and the matter that may be thrown into rivers is regulated by law; in this country some states have also made enactments which in time will no doubt be perfected and enforced. The chemical analysis of water collected below a town which discharges sewage into a river shows a higher proportion of chlorine than is found above the town, and the number of bacteria will also in general be much greater. River water is improved by sedimentation in reservoirs, but if very impure it must be treated by artificial filtration before delivering it to the distributing basins.

Lakes are natural reservoirs which collect the water of brooks and rivers. When of large size they furnish an excellent supply, for both sedimentation and aeration have had opportunity to remove the organic matter. The water of a lake may be polluted, however, by the refuse of towns or by the discharge of sewage into it, so that contamination may extend to a considerable distance from the shores; at Chicago, for instance, the supply is collected in cribs four or five miles from the lake shore and carried in tunnels to the city in order to secure uncontaminated water. In small lakes and ponds the liability to pollution is greater still, and hence filter galleries are frequently used to insure purification.

It should be noted that the quality of water from any given stream or lake undergoes systematic variations with the changes of the seasons. In late winter and early spring the melting ice and snow cause the maximum run-off of the year, and the streams become swollen and foul with organic and inorganic matter. In summer the flow becomes normal and the highest degree of purity obtains; in autumn the flow is a minimum and liability to pollution is greater than in the summer. Even in a pond whose surface does not vary greatly in height there are variations due to the influence of wind and temperature. It was shown by Drown in 1891 that in deep ponds and reservoirs a stagnant layer containing unoxidized organic matter is formed during the summer, and that this rises to the surface late in the fall when the higher layers become cooled; a vertical circulation then occurs with more or less regularity until spring, when the stagnant layer begins to form again. Hence both river and lake waters are liable to be of variable quality from fall until spring, and the necessity for artificial purification is greater then than during the summer season.

20. Ground Waters.

The water of springs and wells comes from that part of the rainfall which has percolated into and through the soil. In the upper layers of the soil a large part of the organic matter is removed by the action of bacteria and absorbed by the growing vegetation. As the percolation extends downward mineral matter is generally taken up, so that ground water is harder than surface water. In good spring or well water the total solids may be as high as 100 parts per million, but very little of this is organic matter; the ammonias and nitrates are very low, while the amount of chlorine should not exceed the normal for the region. If both chlorine and nitrates are found high it is an indication of sewage contamination.

Water rises in springs from the hydrostatic pressure of a body of water which fills the soil below a certain depth. This body of water has a slow motion through the soil towards the streams and the ocean, and it has a more or less definite surface which slopes in the direction of flow. In the sandy soil of Long Island the slope of the surface of this ground water is about 10 feet per mile, but in clayey or rocky soils the slope is not as great. The depth of this water surface below the level of the earth varies in different seasons, and also in years of different rainfall. When the topography of the region is such that this body of ground water is intercepted a spring results and clear water bubbles up at the foot of a hillside.

Well water is merely ground water which is intercepted by sinking a pit into the earth. If the draft be great the surface of the ground water around the well assumes the conoidal shape indicated in the figure, and if it be very great the well

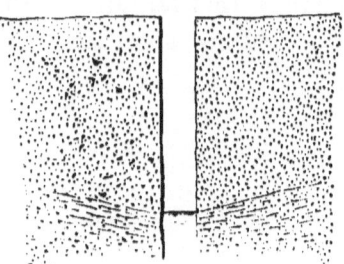

OPEN WELL IN GROUND WATER.

may become dry. When many wells are sunk near together the surface of the ground water becomes permanently lowered and it may be difficult to obtain the supply desired. This has been markedly the case at Brooklyn, N. Y.; at a well sunk in 1869 the ground-water level was lowered five feet in eight years of pumping, so that the construction of other wells became necessary, and the same phenomena continued to prevail at all of these, until finally many wells at considerable distances from the city are required to maintain the supply.

Driven wells are pointed iron tubes with holes near the lower ends which are sunk into the soil until they pierce the ground-water level. The water is raised by a suction-pump at the top of the tube, or sometimes several tubes are connected to one pump; of course, water cannot be raised by this method higher than about 30 feet above the ground-water level. There is nothing mysterious in a driven well, as many suppose, nor can any more water be obtained from it than from an open one by the same expenditure of work.

House wells are liable to become polluted by the drainage of outhouses and cesspools; the older the well the greater is the danger, for the surrounding soil is apt to become fouler year by year, and finally, when an unusual lowering and subsequent inflow of ground water occurs, contamination results. In numerous cases boards of health have found it necessary to order wells to be filled up on account of the disease and death which they have caused.

Deep wells sunk from 100 to 300 feet in rock furnish water for towns as also for houses and manufactories. Here the ground water pervades the fissures of the rock and is intercepted by the well; it is usually raised to the surface by a force or lift pump, which is operated by a plunger from an engine on the surface. The quality of such water does not differ essentially from that of the surrounding ground water, except in the larger proportion of mineral matter.

Artesian water comes from a well carried down to pierce a water-bearing stratum whose outcrop is at a considerable distance from the well. Thus in the figure the water falling on

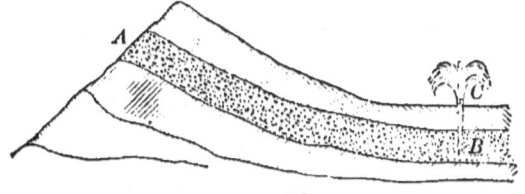

ARTESIAN WELL.

the outcrop A is confined by the geologic conditions to the stratum AB, and rises through the well BC by virtue of the hydrostatic pressure. In some cases the water spouts up above the top of the well, while in others pumping is necessary. Many wells of this kind have been sunk in the United States; one at Charleston, S. C., is 1260 feet deep, and its water level rises and falls with the tide, although not simultaneously; at St. Louis, Mo., there are two whose depths are 2199 and 3843 feet. Artesian waters are generally so highly impregnated with metallic salts as to be unsuited for domestic uses, the total solids being often from 500 to 1000 parts per million. The chlorine, ammonias, and nitrates are also high, but these do not indicate contamination, as in surface and ground waters. The main use of artesian water is in manufactories, where hardness and mineral salts are not objectionable, but in some cases it is also collected for the supply of towns.

21. Reservoirs.

The surface water of swamps and brooks and the ground water of springs is collected in reservoirs for storage and distribution. River and lake water is also pumped into reservoirs and tanks, or it is often delivered directly into the pipes without the use of reservoirs. The use of reservoirs not only enables a sufficient supply to be maintained, but also affords opportunity for clarifying and purifying the water.

Reservoirs for irrigation were extensively built in Egypt and India in ancient times. Lake Maeris in Egypt, constructed about 2000 B.C., is said to have been 413 miles in circumference. The Romans built reservoirs for supplying water to their aqueducts, selecting the water of springs in preference to surface water. In India there are thousands of reservoirs still in operation which were constructed in early times for purposes of irrigation; one of the modern reservoirs has an earthen dam 12 700 feet long and a storage area of 4.2 square miles.

Collecting and storage reservoirs are those that impound the water of streams, while distributing reservoirs are those whose water is supplied either from storage reservoirs or by pumping from rivers and lakes. From the distributing reservoir the water goes to the town through mains, under the action of gravity; in some cases, however, the same reservoir may serve both for storage and for distribution. A collecting reservoir is usually made by constructing a dam across a valley, thus creating an artificial lake, while distributing reservoirs are made by forming an excavation on a hill. The engineering features of these constructions are to be described in the next chapter, while here those points are to be discussed which are important for insuring the purity of the water.

A collecting reservoir formed by throwing a dam across a valley impounds the run-off of the water shed above that dam. It is very necessary that the area to be covered by the impounded water should be carefully cleaned; all trees and bushes should be removed, and if the upper layer of the soil contains vegetable matter it should also be excavated and carted away. Neglect of these precautions has frequently caused the water to be of bad quality, on account of the absorption of decaying organic matter. For instance, at Wilkesbarre, Pa., a swampy region was in 1891 turned into a storage reservoir by the construction of a dam and cleaning was imperfectly done; the poor quality of the water rendered an extensive filtration plant necessary, but even this could not satisfactorily clarify the water, so that later the entire system was abandoned and a supply for the city obtained from another source. For purposes of irrigation alone the cleaning of a reservoir area is not necessary, as the organic matter in the water is beneficial rather than otherwise to growing crops, but for all public water supplies such cleaning is indispensable.

The bed of a storage reservoir should hence, if possible, be formed of clean gravelly earth or of rock. Its banks should be free from all rubbish that is liable to be washed into the water in times of storm, and a certain degree of sanitary patrol should be established around it. In most states the law provides for the proper punishment of persons who pollute a reservoir by throwing organic matter into it or by bathing in its water, but the enforcement of these laws is far from universal. Moreover, unintentional pollution is liable to occur from farmhouses and barns situated on the watershed, and this is to be prevented only by a regular system of inspection. Probably the most complete system of sanitary patrol of a reservoir in the United States is that exercised at Hemlock Lake, which supplies Rochester, N. Y., the refuse of all farmhouses around the lake being systematically carried away and every precaution taken to prevent contamination.

A distributing reservoir is smaller than a collecting one, and usually its bottom and sides are made of concrete, in order to prevent leakage. The mains enter it in a gate-house, which is often so arranged that the water may be taken from different depths below the surface. There is also a waste pipe running out from the bottom, which is used to entirely empty the reservoir and to remove the mud deposit that may have formed in it. A similar method of cleaning out the deposits in a collecting reservoir may also be used, provided that the supply of the town is not impaired by emptying it. Organic matter in a distributing reservoir is far more dangerous than in a collecting one, and hence the access of the public to it should be absolutely prevented by means of an effective enclosure.

The quality of water in a reservoir is always improved by putting a roof over it, but this is only practicable when the area is small. The roof protects the water from sunlight, and hence renders its temperature more uniform and retards the growth of diatoms and other vegetable organisms. As indirect advantages the evaporation is lessened and pollution by birds is prevented. At Paris there is a covered reservoir whose size is 136×272 meters, at London there are two, and at Naples there is an underground reservoir in rock galleries on a hill.

Analyses of reservoir water, made monthly or at shorter intervals, are of great value in giving knowledge of the changes that go on with the progress of the seasons, in giving warnings of danger, and in showing the effect of improvements in methods of collection. A single isolated analysis, on the other hand, is of little value, as it gives no opportunity for comparisons.

22. SEDIMENTATION AND AERATION.

Sedimentation and aeration are two processes by which nature purifies water. Sedimentation or settling occurs when a body of water is kept quiet, so that the suspended matter may slowly settle. Aeration is bringing the water into contact with air, so that its oxygen may enable the bacteria to decompose both the suspended and dissolved organic matter. These processes must be carried on at different times, since aeration involves the agitation of the water.

Sedimentation is consequently going on in all collecting and distributing reservoirs. The force of gravity acting on the suspended organic and inorganic matter causes it to slowly fall, provided that its specific gravity is greater than that of the water. The downward force on any particle is the difference between its weight and the weight of an equal volume of water, and it is clear that this may be very small for organic matter, and that hence the velocity of fall is also small. Moreover, the theory of the descent of a small spherical body in a resisting medium shows that the velocity soon becomes uniform and that it varies as the square root of the diameter of the body. Accordingly, very small particles of silt fall very slowly in water and take a long time to reach the bottom of the reservoir. So fine is the division of the silt in the water of the Missouri River that bottles of it have been kept for years without becoming fully clarified. To secure the quickest rate of sedimentation it is plain that the water should be very quiet and that its depth should not be too great.

The process of sedimentation consists mainly in the fall of

the suspended inorganic matter, but this in descending also drags down some of the suspended organic matter with it. As the organic matter is closely of the same specific gravity as water, the velocity of fall of a grain of silt, whose specific gravity may be double that of water, is retarded on striking a particle of organic matter, and thus another reason is apparent why the process of sedimentation is so slow. But if sufficient time be afforded a large proportion of the suspended inorganic matter will ultimately be precipitated, with the result of rendering the water clearer and more suitable for domestic uses.

Aeration consists in causing the water to dissolve oxygen out of the air. This oxygen immediately attacks both the suspended and dissolved organic matter in the manner explained in Art. 6, and the process of nitrification proceeds until the ammonia has disappeared by being transformed into nitrates. This process goes on to a greater or less extent in all impure water to which sufficient oxygen is supplied, provided such water contains in solution potash, soda, and other metallic alkalies. Sedimentation purifies the water by removing suspended matter, and a large part of this is inorganic matter only, but aeration purifies it by removing both the suspended and the dissolved organic matter.

A large collecting reservoir will have its surface agitated at times by winds, and thus considerable air is absorbed by the upper layers of the water. In transferring the water from a collecting to a distributing reservoir there is frequently found the opportunity to give it more satisfactory aeration. If the transfer be made by an open canal, this can be built so that there may be one or two falls where the water flows over rocks and is broken up into foam. If a water main be used, the terminus at the distributing reservoir can be made so as to throw the water up in fountain spray and thus secure a thorough admixture of air. The experience at Reading, Pa., in 1880–85, may be cited as showing the advantage of the second method. A portion of the storage at Antietam Lake was directly distributed to the city, while the remainder was carried to distributing reservoirs and entered them by fountains which threw the water to a height of 75 feet. In the part of the city served directly from the lake disagreeable odors and tastes were observed in August of each year, but in the sections served from the distributing reservoirs the water was entirely free from such unpleasantness. The offensive odor and taste were doubtless due to the decay of diatoms or other vegetable growths, but the aeration at the distributing reservoirs had completely oxidized the decaying matter.

The following chemical analyses will give a general idea of the changes that occur in brook water by storage and sedimentation. The reservoir in question is one at Springfield, Mass., and the analyses are the mean of several made by the Massachusetts State Board of Health in April and May, 1889. The samples of water from the brooks were collected where they entered the reservoir, and those from the reservoir itself were collected near the middle and at a depth of six feet below the surface. The figures express parts per million:

	Brooks.	Reservoir.
Total Solids,	33.9	23.8
Organic Matter,	13.9	9.5
Inorganic Matter,	20.0	14.3
Chlorine,	0.8	1.0
Free Ammonia,	0.017	0.002
Albumenoid Ammonia,	0.163	0.190
Nitrogen as Nitrates,	0.023	0.023
Nitrogen as Nitrites,	0.004	0.003

Here the total solids, both organic and inorganic, are diminished, and the general increase in purity is indicated by the decrease in free ammonia. This reservoir, it may be noted, has always given more or less trouble by odors due to decaying algæ, and it may be that the presence of living algæ prevented the increase in nitrates which would naturally be expected.

23. Natural Filtration.

Sedimentation, aeration, and filtration are the three methods by which nature purifies water. The word filtration means primitively screening or straining, but when applied to water purification it has a much more extended signification. It means not only the removal of the suspended matter by screening or straining, but the removal of the dissolved organic matter by nitrification which is produced by bacteria in the presence of oxygen. In the article on ground waters a general description of this process was given, but it is well to consider it in fuller detail, as the theory is of great importance in the discussion of methods of artificial filtration.

The rainfall which percolates into the earth collects impurities from the surface, some of which are screened out in passing through the upper layer of the soil, while others remain in solution. The interstices of the soil are filled with air which has entered prior to the fall of rain, and the water thus becomes charged with dissolved oxygen. Each particle of sand or earth is surrounded by a film of water which is subject to a slow downward motion from the pressure above it, and in these thin films the bacteria work with great activity in decomposing the organic matter. As the water descends in the soil more and more organic matter becomes nitrified, but the process ceases after a depth of four or five feet is reached, because the difficulty of getting air prevents the bacteria from carrying on their useful work. The descent of the water continues, however, until the level of the ground water is reached, and in this body of ground water more or less sedimentation goes on, which removes some of the suspended matter still remaining, while the dissolved inorganic matter increases by absorption of metallic salts from the earth. This is the process by which impure surface water is converted into the clear water of springs and wells.

A water supply derived from ground water has hence been purified by natural filtration. The amount of chlorine in the surface water is not decreased by the filtration, but if an increase is observed this is derived from the soil. If the normal chlorine in a spring or well water is found to increase, this is almost certainly due to pollution from sewage, as all sewage contains salt which is derived from the food of men.

Public water supplies of ground water may be obtained from open or driven wells, as explained in Art. 20, and also by means of filter galleries or tunnels. The figure shows a filter gallery in the ground which is covered with an arch and has an open bottom through which the ground water enters; in water-bearing rock an artificial arch may not be required. The city of Dubuque, Ia., derives its supply from a filter gallery a mile long driven horizontally into rock at a depth of more than 100 feet below the surface of a bluff.

FILTER GALLERY.

Filter galleries are also constructed along the banks of rivers, so that the river water may percolate into them and become purified in its passage through the soil, but in such cases a part of the supply is apt to be ground water; such galleries have been built at Brookline, Mass., Columbus, O., and many other places. A filter basin is similar in principle to a filter gallery, but it usually has an open top and is thus merely a very large well. The object of all such constructions is to collect water which has been purified by the method of natural filtration.

The following figures, which give average results of a large number of chemical analyses made by the State Board of Health of Massachusetts in 1887-89, show the effect of natural filtration at Framingham, Mass. The source of supply is a filter gallery which receives its water from a pond, part of the gallery being underneath the pond and the water entering through the bottom, as shown in the preceding figure; in order to enter the gallery the pond water must pass through at least five feet of earth. All figures are, as usual, expressed in parts per million:

	Pond.	Filter Gallery.
Hardness,	16	30
Total Solids,	51.0	59.3
Organic Matter,	16.3	9.0
Inorganic Matter,	34.7	50.3
Chlorine,	4.0	4.5
Free Ammonia,	0.047	0.031
Albumenoid Ammonia,	0.262	0.084
Nitrogen as Nitrates,	0.158	0.272
Nitrogen as Nitrites,	0.003	0.003

Here the effect of the purification is seen in the decrease of the organic matter and ammonias and in the increase of the nitrates. The increase in hardness and chlorine indicates either that some ground water reached the gallery or that the filtering soil contained calcium carbonate and sodium chloride. The permanence of the nitrites shows that the purification was not fully satisfactory.

24. Artificial Methods of Purification.

The three methods by which nature purifies water are sedimentation, aeration, and filtration, and in the preceding articles it has been shown how these are used by man in allowing subsidence in reservoirs, in causing air to be dissolved by the water, and in collecting ground water which has been filtered by passage through the earth. There are, however, other methods used by man; a brief classification of these will here be given, and the most important of them will receive fuller discussion in subsequent articles.

Distillation of water consists in turning it into steam, which is again condensed into water. This is undoubtedly the most effective process of purification, but it is so expensive that it is only to be used in exceptional cases, as on steamships; most of the vessels of the United States navy are provided with apparatus for distilling salt water. The heat destroys the bacteria, and the evaporation leaves both suspended and dissolved solid matter behind, with the result that distilled water is the purest obtainable, although its lack of taste renders it somewhat unpalatable.

Boiling is an effective method for the purification of water if it be continued for half an hour or more, as the bacteria are killed by the heat and hence the organic matter is rendered harmless. It is said that the boiling of drinking water is a widespread custom in China and Japan, and it may perhaps be partly due to this that these countries are free from the frightful epidemics which are so common in India. Boiling is also an expensive process, and cannot be carried out except in a household where fire is needed for other purposes.

Hard water frequently requires to be softened before it can be economically used for washing, and in steam-boilers its use is objectionable on account of the carbonate of lime which is deposited. Clark's process for softening water of temporary hardness is to add limewater (CaO, H_2O); this unites with the free carbon dioxide (CO_2) in the water to form calcium carbonate ($CaCO_3$), which is immediately precipitated and drags down with it the calcium carbonate in solution. If the hardness be permanent, or due to calcium sulphate ($CaSO_4$), it is necessary to add sodium carbonate (Na_2CO_3); the combination of these produces calcium carbonate ($CaCO_3$) and sodium sulphate (Na_2SO_4), and the former of these is precipitated. In England many public water supplies have such a high degree of temporary hardness that the limewater process is used on a large scale. In America, however, a public water supply is rarely so hard as to require such treatment, and in households sodium carbonate is generally used for softening waters both of temporary and permanent hardness.

Chemical precipitation is also used to clarify water, particularly in connection with methods of mechanical filtration. Electrical methods of purification may be classed with the chemical ones, and both will be discussed in Art. 25.

Household filters depend for their efficiency mostly upon the screening or straining action, and do not use the true filtration principle of removing organic matter by nitrification. Some of these will be described in Art. 26.

Mechanical filters are those that force water through sand or other earthy material; these employ the nitrification principle only to a slight extent. In Art. 27 one or two of the most common kinds will be explained.

Artificial filtration through beds of sand, carried on slowly and intermittently, so as to imitate the process of natural filtration, is the most effective method of purifying an unsatisfactory public water supply. The construction and operation of such filter beds will be discussed and illustrated in Arts. 28 and 29.

25. Chemical and Electrical Methods.

All chemical treatment which causes precipitation also purifies the water of suspended organic matter. Thus, the addition of limewater to water containing calcium carbonate causes this to be precipitated, and it, in descending, drags down a considerable proportion of the suspended matter. Most of the common chemical methods of purifying water depend upon this principle.

Alum is one of the best-known precipitants. This is a double salt of aluminum and potassium sulphates $(Al_2(SO_4)_3 . K_2SO_4)$, and when added to water (H_2O) containing calcium carbonate $(CaCO_3)$ in solution there results carbon dioxide (CO_2), calcium sulphate $(CaSO_4)$, potassium sulphate (K_2SO_4), and aluminum hydroxide $(Al(OH)_3)$; thus

$Al_2(SO_4)_3 \cdot K_2SO_4 + 3H_2O + 3CaCO_3 = 3CO_2 + 3CaSO_4$
$+ K_2SO_4 + 2Al(OH)_3.$

Here the carbon dioxide is a gas, the calcium and potassium sulphates remain in solution, but the aluminum hydroxide is precipitated in the form of a flocculent white salt which in descending clears the water of the suspended matter, both organic and inorganic. If the water does not contain calcium carbonate there must be added sodium carbonate (Na_2CO_3), and the reaction is the same as before, Na taking the place of K, while the aluminum hydroxide is precipitated. On account of the cheapness of alum it is extensively used in the purification both of water and sewage.

Iron perchloride ($FeCl_3$) and sodium carbonate (Na_2CO_3) added to water (H_2O) produce carbon dioxide (CO_2), sodium chloride (NaCl), and iron hydroxide ($Fe(OH)_3$), the last being a substance which precipitates and clears the water. This method has not come into use.

Potassium permanganate ($K_2Mn_2O_8$) added to water gives a purple color, and if used in sufficient quantity so that the color persists for ten minutes it causes effective purification. This is caused partly by the liberation of oxygen, but mainly by the formation of a manganese hydrate which is precipitated. This process is an expensive one on a large scale and has been used but little. A solution of alum and potassium permanganate has been recommended to be used by soldiers to purify and clarify foul water which is found near their camps.

The so-called electrical methods of purification are really chemical ones, as the electric action merely causes chemical reactions to take place. One of these is the decomposition of some of the water into oxygen and hydrogen, with the intention of furnishing free oxygen to act upon the organic matter like the oxygen furnished by aeration. In fact, however, the oxygen generally combines with the iron plates at the positive pole, and the iron oxide thus formed is precipitated and clarifies the water. This method can be employed only on a small scale, and its use is very limited. When aluminum plates are used at the positive pole an aluminum hydrate is precipitated.

A second electrical method is to make a 2 per cent solution of common salt in water, pass the electric current through it, and then add about one gallon of the resulting liquid to 6000 gallons of the water to be purified. The effect of the electric current on the sodium chloride ($NaCl$) is to liberate the chlorine (Cl), which in connection with the free oxygen in the water forms sodium hypochlorite ($NaOCl$). This substance is a poison which kills the bacteria and thus prevents the propagation of disease as far as these are immediately concerned, but the action is no different from that of sodium hypochlorite prepared by other methods. The addition of a disinfecting poison to water is, however, not regarded with favor, and consequently this method has not come into use.

26. Screens and Strainers.

All household filters are arranged so as to screen or strain the water as it passes through them. A fine sieve or screen may remove most of the suspended matter, but the dissolved matter will pass through it; moreover, after a few days of use the screen becomes so clogged with the suspended matter that some of this may be dissolved out and thus render the water more impure than before the operation. All household filters hence require frequent cleaning in order to maintain their efficiency.

Charcoal, and particularly charcoal made from the bones of animals, is one of the most effective strainers. It is formed into plates, and these are arranged in a box so that all the water issuing from a pipe is compelled to pass through them by the hydrostatic pressure. The surfaces of the plates, and their interstices also, soon become clogged with organic matter, and it is necessary to take them out and remove the organic matter by heating in order that the purification of the water may continue. In an experiment by Frankland it was found that no bacteria appeared in the filtered water during the first twelve days of use of the charcoal; at the end of a month, however, the filtered water contained 7000 bacteria per cubic centimeter, which was five times as many as were found in the unfiltered water. In general, the use of these charcoal screens is a source of danger rather than a benefit.

The Pasteur filter consists of a porcelain cylinder contained within an iron tube with an annular space between; the iron tube is connected to the house-pipe and the water fills the annular space and is forced through the porcelain under the hydrostatic pressure when the faucet is open which connects with the inner cylinder. At first this completely removes the bacteria, but after a week or two they are found in the filtered water, having passed through the porcelain by virtue of their multiplying numbers. To secure pure water it is necessary that the porcelain cylinder should be removed once a week and be thoroughly boiled, to kill the bacteria that it contains.

On a large scale, coke may be used as a strainer, but this is done more commonly for sewage than for water. At intervals of a few days fresh coke must be supplied, while the old is taken out and heated in ovens in order to burn the organic matter and kill the bacteria.

Spongy iron, made by blowing air through iron ore in a highly heated state, is a porous material which has been much used in Europe. It may act as a strainer, but its true action is that of causing precipitation, which it does by the production of a ferric hydroxide under the action of the free oxygen in the water.

It is seen that these screens and strainers do not use the

principle of purifying water by natural filtration which has been explained in Art. 23, as that method removes the organic matter through a slow process under the action of bacteria and oxygen. Although popularly called filters, they are so only in the sense of sieves or strainers, and the principle of their action should not be confounded with that of true filtration.

27. Mechanical Filters.

A mechanical filter is one that is operated by power and strains the water at a rapid rate through sand or other suitable material, afterwards discharging it into a basin, where sedimentation may take place. Alum is often employed as a precipitant, this being added in the form of a solution before the water enters the filter. Mechanical filters are frequently used in hotels and hospitals, as also for the public supplies of towns and small cities.

The mechanical filters used in America are quite similar in principle, although they differ much in detail and arrangement. In some the water passes through under the action of gravity, while in others it is forced through by pumps, the former having wooden tanks with an open top, while the latter are made of iron or steel and are closed. Most of them are vertical cylinders, the water entering at the top and coming out at the bottom, but some are cylinders placed in a horizontal position. Some companies employ alum, while others do not, the use of this in a mechanical filter being controlled by a patent. A general description of the method of operation of a gravity tank will render clear the principle of action of all American mechanical filters.

Imagine a cylindrical wooden tank, say 12 feet in diameter and 20 feet high. The water is brought into this tank through a pipe A, and by means of an apron is scattered so that it may receive some aeration. Near the bottom of the tank is seen a sand-bed, B, about two feet thick, contained in an iron box with a perforated bottom. The pressure due to the head above B forces the water through this sand to the chamber C, whence it passes through the pipe D to the settling basin. After a time, varying from six to twenty-four hours, according to the degree of impurity of the water under treatment, the sand-bed becomes clogged with dirt and the rate of straining is so slow that the process is stopped in order

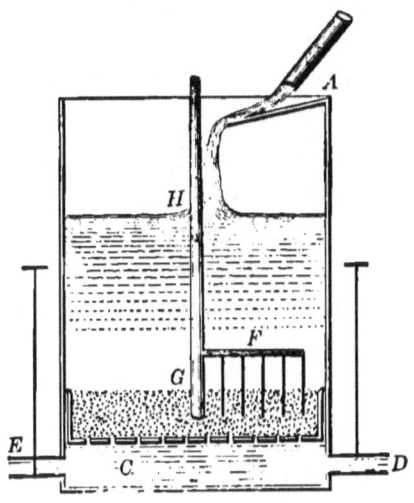

MECHANICAL GRAVITY FILTER.

to clean and wash the sand. For this purpose the valve at D is closed and the chamber and tank are drained by a pipe not shown in the figure. Then the valve E is opened and water is forced by a pump into the chamber and upwards through the sand-bed; simultaneously the rake F is lowered into the bed and revolved by the axle GH, in order to thoroughly stir the sand. This dirty water is then drained off, and the cleaned bed is ready for the next operation.

While the above description does not exactly represent any particular mechanical filter, yet it is believed that it gives a fair account of the principles involved in all of them and of the general method of operation. A filter of the size stated generally delivers from 300 000 to 400 000 gallons in twenty-four hours, or approximately at the rate of 3000 gallons per square foot of surface per day. When arranged for the purification of a public supply a number of them are required, the pumps and machinery for all being driven by the same engine. There are in the market about six principal styles, each having special patents covering the details of arrangement and methods of washing. In all cases the principle is that of straining, supplemented by frequent cleaning of the strainer, and the natural process of removal of organic matter by nitrification is imitated but slightly.

The following analyses show the result of the purification of water at Long Branch, N. J., in 1892, by one of these mechanical filter plants, the figures giving parts per million. This water was from a peaty swamp and had a dark yellow color, which was entirely removed by passage through the filter.

	Before.	After.
Hardness,	22.5	22.5
Total Solids,	95.2	71.4
Organic Matter,	42.4	17.4
Inorganic Matter,	52.8	54.0
Chlorine,	3.5	3.5
Free Ammonia,	1.320	0.035
Albumenoid Ammonia,	0.445	0.095
Nitrogen as Nitrates,	0.876	0.870
Nitrogen as Nitrites,	0.025	0.001
Oxygen consumed,	12.3	1.9

There can be no doubt but that the mechanical sand filters are able to purify all but the most incorrigible waters if the cleaning be carried on at intervals sufficiently frequent. The construction and operation of such a mechanical plant is a matter that involves considerable expense to a town, but the same must be said regarding the system of artificial filtration by sand-beds which is to be described in the following articles. Which of these should be selected for the treatment of an unsatisfactory supply can only be decided after a comparison of plans prepared by an experienced engineer.

Anderson's iron method of purification is mainly used in Europe as an auxiliary process to filtration through sand-beds. An iron cylinder is arranged with fixed vanes or ledges projecting from its interior surface and into it iron shavings are put. The cylinder revolves on hollow bearings at its ends, through which the water enters and leaves, and being revolved, the iron shavings are thrown about by the vanes and brought into contact with the water. The result is that an iron hydrate is formed, and this is allowed to precipitate as the water flows through troughs or basins on its way to the sand filtering beds. It is stated that one cylinder is able to purify about 600 000 gallons of water per day at a cost of about 50 cents when a plant operating a number of cylinders is installed.

28. Artificial Filtration.

Artificial filtration is an imitation of the process of natural filtration which is described in Art. 23. Beds of sand and gravel are prepared on the ground, and the water is allowed to pass through them at a slow rate, so as to afford sufficient time for the useful bacteria to decompose the organic matter into harmless constituents. The purified water, often called the effluent or the filtrate, runs out at the bottom of the beds and is collected in basins for distribution. In this manner an impure surface water is turned into a pure ground water, the organic matter and ammonias being decreased, the chlorine remaining constant, and the inorganic matter and nitrates being increased.

The following figure shows a vertical section of a filter bed. At the top there is about two feet of fine sand, below this about one foot of gravel, and then about one foot of broken stone. From this broken stone the water passes through holes in the wall to a basin, from which it is pumped or delivered by gravity to the town. The filter bed has a con-

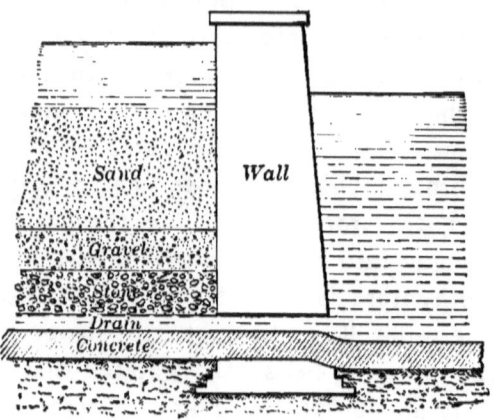

SAND FILTER BED.

crete bottom, in order to prevent the inflow of ground water and to insure that the effluent shall pass through the drains into the receiving basin.

When this method is to be applied to a river water it is first pumped to a reservoir which affords opportunity for sedimentation. In the passage from the reservoir to the filter bed it is well also to cause aeration in order to furnish oxygen to the bacteria. The proper rate of flow through the bed is insured by regulating the height of the water level over the beds or in the receiving basin so that the head may be sufficient to cause that flow. Different kinds of water require different rates of filtration; common river water needs about one square foot of filter-bed surface for each 60 gallons of water filtered in one day, or the rate of filtration is 60 gallons per square foot of surface per day. A very impure river water may, however, need a rate as low as 30 gallons per square foot of surface per day, while a lake water may be filtered at a rate as high as 100 gallons per square foot of surface per day. The more impure the water the lower must be the rate of filtration, as more time is required for the bacteria to decompose and nitrify the organic matter.

The depths of the different layers of the filter bed and the relative proportions of sand and gravel used are subject to much variation in constructions by different engineers. It is, however, universally agreed that the upper sand layer is the most important, since the activity of the bacteria is the greatest near the surface, where fresh air is always present. It is for this reason that the rate of filtration is practically independent of the depth of the filter bed. One foot of fine sand in the upper layer is probably as efficient as two feet, but owing to the diminution in depth caused by cleaning the original depth of this should be greater than one foot. The intermediate gravel layer acts merely as a support to the sand, while the stone layer serves to distribute the water to the underdrains which are laid at intervals through it.

This method of artificial filtration was developed in Europe about 1860 as a result of the study of natural filtration. It is now extensively used there, and furnishes purified water to more than 20 000 000 people. In Art. 9 one of the instances is given where the use of such filter beds prevented the spread of cholera, and everywhere it has been found that the rate of deaths from typhoid fever has been materially decreased. In the United States the method has been used at Hudson, N. Y., Poughkeepsie, N. Y., and Lawrence, Mass., and a few other places; at Lawrence, the typhoid death rate was reduced nearly one-half in the first year of its use. The development of water supplies in America since 1870 has been so rapid that the question of an abundance of water has received more attention than that of purity, but the time has now come when filtration through sand-beds is receiving earnest attention, and a number of cities have prepared plans for its introduction.

The following analyses of the water from the Merrimac River, and of the same water after passing through the Lawrence filter beds, will give an idea of the results accomplished by artificial filtration; the figures are the averages of daily analyses made by the Massachusetts State Board of Health during the month of July, 1895:

		River.	Effluent.	Reservoir.
Hardness,	parts per million	18	24	24
Chlorine,	" "	2.7	3.9	2.8
Free Ammonia,	" "	0.134	0.075	0.022
Albumenoid Ammonia,	" "	0.243	0.097	0.099
Nitrogen as Nitrates,	" "	0.110	0.450	0.450
Nitrogen as Nitrites,	" "	0.003	0.001	0.001
Oxygen consumed,	" "	3.8	2.4	2.1
Bacteria per cubic centimeter,		10 000	50	69

Here the purification of the effluent is shown by the decrease in the ammonias and nitrites and by the increase in nitrates, but particularly by the decrease in the number of bacteria. The effect on the effluent by subsequent sedimentation in the distributing reservoir is mostly apparent in the decrease of free ammonia and oxygen consumed, but the July heat caused a slight increase in the number of bacteria.

29. Operation of Filter Beds.

The size of a filter area depends upon the quantity of water to be filtered and the rate of filtration. For a town of 25 000 inhabitants, using 100 gallons per person per day, the average consumption will be 2 500 000 gallons per day, and at a rate of filtration of 60 gallons per square foot of surface per day about 42 000 square feet, or nearly one acre, of ground is required. Owing to the cleaning of the surface, which is periodically necessary, and owing to the fact that the maximum consumption may be much greater than the mean, it is well to provide a larger area, say $1\frac{3}{4}$ or 2 acres. This may be divided into three or four beds, so that one may be thrown out of use when it becomes necessary to clean its surface.

Each bed is separated from the others by walls and its bottom is water-tight. A series of drains with loose joints is laid on the bottom, and these are connected with one or two main drains which discharge the effluent. Arrangement must be made to waste the effluent instead of allowing it to run to the settling basin, as this is necessary for a day or two after starting the filtration; also means for draining the bed when it is to be put out of use must be provided. The stone, gravel, and fine sand are then applied in successive layers, great care being exercised that at the same depth below the surface the distribution should be uniform over the entire area of the bed. The sand in the upper layer should be so fine that ten per cent of it has grains whose diameter is between 0.2 and 0.4 millimeters.

When the water is admitted upon the filter bed the rate of flow will be greater for the same head than after it has been in operation for a few days. This retardation is due to the collection of organic matter in the top of the sand, forming a so-called dirt layer, from one-half an inch to one inch in thickness, and it is found that effective purification does not occur until this has been formed. Hence for the first day or two it is best to waste the effluent instead of collecting it. This dirt layer should not be broken, for if so the water that passes through the holes is ineffectively purified. As the dirt layer becomes thicker and thicker the flow becomes more and more impeded, so that finally, after an operation of from two to six weeks, it is found necessary to drain the bed and clean its surface.

The cleaning is accomplished by removing that part of the sand which contains the dirt layer, the thickness of this being usually about one inch. The filter bed is then put into operation again until a second cleaning is necessary. After several inches of sand have been removed fresh sand is applied to restore the thickness, the dirty sand being generally washed for this purpose.

The great importance of the dirt layer will be better appreciated when it is stated that it contains more than one-half of the total number of bacteria in the filter bed; at a depth of one-quarter inch below the surface the number of bacteria is about ten times as great as at a depth of one inch, and at a depth of two inches the number is only about one-fourth of that at a depth of one inch. These bacteria are doing the useful work of decomposing the organic matter of the water by permitting the oxygen to cause its combustion and nitrification.

The method above described is called continuous filtration, as the head of water remains constant in the interval between the cleanings. Another method, which is less often used, is that called intermittent filtration; in this the bed is drained from time to time, in order to allow the air to enter and thus furnish oxygen to the bacteria. The filter beds at Lawrence, Mass., are of this type, they being drained once a day to secure aeration, and the cleaning, when necessary, is done during the periods of daily rest. It is still an undecided question as to whether the results of the intermittent method are better than those of the usual continuous method.

In Europe many filter beds are covered with masonry vaults to protect them from the action of frost, as this impedes the activity of the bacteria and hence lessens the efficiency of the purification. The cleaning of an open filter cannot be well done in freezing weather, while that of a covered one is done as effectively in winter as in summer. The cost of a masonry roof adds, however, greatly to the expense of construction.

The cost of construction and maintenance of a filter-bed system constitutes a material addition to the usual expenses of operating a water-supply system. An estimate by Hazen for a city using 10 000 000 gallons per day and having an area of five acres for the filter beds gives $350 000 for the cost of construction if the beds are to be vaulted, and about $43 per day as the cost of operation. Taking into account the interest on the cost of construction and the sinking-fund contribution necessary to repay the same, the total cost of filtration may be put at $1\frac{1}{4}$ cents per thousand gallons, or about 46 cents per person per year, if the average consumption be reckoned at 100 gallons per day. But this expense must be met when a city is using a water so impure as to raise the typhoid death rate above the normal for the surrounding region.

In conclusion it may be said that although artificial filtration is an imitation of the process of natural filtration, it differs from it in one particular. In nature the process of purifying surface water is materially aided by the vegetation growing on the surface, as this absorbs not only the most impure water but also the products of nitrification. In artificial filtration the office of vegetation is replaced by the process of cleaning the filter beds, and the manner in which this is done is most important. In fact, to secure the best results it is indispensable that regular bacteriological examinations of the unfiltered water and of the effluent should be made, in order that the effect of the methods of operating and cleaning the beds may be definitely known, and that warnings may be given of any imperfections. In Germany such bacteriological analyses are required to be made daily. Frequent chemical analyses are also of value to supplement and verify the conclusions of the bacteriological ones, and detailed records of the rate of filtration must be kept. Thus by well-laid plans and eternal vigilance in executing them the sanitary engineer transforms the water of a foul river into a public supply of the highest degree of purity.

30. Exercises and Problems.

16 (*a*) Consult Report of United States Weather Bureau for 1891-92, page 32, and describe the Eccard self-recording rain and snow gage.

16 (*b*) Consult Science for December 2, 1892, and state views regarding the influence of the moon on the rainfall.

17 (*a*) Consult Fitzgerald's article in Transactions of American Society of Civil Engineers for 1886, and describe his self-recording evaporometer.

17 (*b*) Consult Transactions of American Society of Civil Engineers, Vol. XXVII, p. 258, and draw a curve showing how evaporation in Massachusetts varies during the different months of the year.

18. Consult Mason's Water Supply (New York, 1896), and interpret the chemical analyses of city and country snow on page 213.

19. Consult Report of State Board of Health of Massachusetts for 1891, and describe more fully the investigations of Drown on the semi-annual turnover of the water in deep ponds and reservoirs.

20 (*a*) Consult the same report for 1892, pages 715-725, and describe how a typhoid fever epidemic in Springfield, Mass., was caused by milk which had been infected from a polluted well.

20 (*b*) What is the derivation of the word Artesian? Describe the artesian well at Grenelle, France; also those of Terre Haute, Ind., Columbus, O., and Chicago, Ill.

21 (*a*) Read the description of Lake Mœris given by Herodotus. What methods did Hippocrates advise for the purification of drinking water?

21 (*b*) Consult Baker's Manual of American Water Works for 1888, and ascertain the capacities of the reservoirs at Reading, Pa.

22 (*a*) An approximate formula for the velocity of fall of a smooth spherical body in water is $v = \sqrt{2gd(s-1)}$, in which g is the acceleration of gravity, or 32.16 feet per second, d the diameter of the body, and s its specific gravity. Compute the velocity v for a particle having a diameter of one-millionth of a foot and a specific gravity of 1.2.

22 (*b*) Consult Engineering Record, August 13, 1898, and give

facts regarding the reduction of number of bacteria by sedimentation.

23. Consult Report on Water Supply and Sewerage of Massachusetts State Board of Health (Boston, 1890), and give descriptions of filter galleries and basins at Framingham, Newton, Waltham, and Wellesley.

24 (*a*) One of the seven articles that a Buddhist monk is allowed to possess is a sieve. What use does he make of it, and why?

24 (*b*) Explain what occurs when sodium sulphate (Na_2SO_4) is added to water containing calcium carbonate ($Ca.CO_3$) in solution.

25 (*a*) If 100 pounds of commercial alum ($Al_2(SO_4)_3.K_2SO_4.24H_2O$) be dissolved in water, show that the weight of the precipitate is about 16 pounds.

25 (*b*) Consult Drown's article in Journal of New England Water Works Association for 1894 (also in Engineering News, March 22, 1894), and describe more fully the electrical methods of purification.

26. See Nichols' Water Supply (New York, 1883), and make a sketch showing a filter of spongy iron for household use.

27. Visit a mechanical filter plant and write a report descriptive of its operation. Collect samples of the original and effluent water and test them by potassium permanganate.

28 (*a*) Consult Report of Massachusetts State Board of Health for 1895, pp. 571-581, and compare the analyses of the Lawrence water during January with those for July.

28 (*b*) Consult Hazen's Filtration of Public Water Supplies (New York, 1896), and describe how the efficiency of filtration depends upon the size of the sand grains in the upper layer of the filter bed.

29 (*a*) Consult Hill's Purification of Public Water Supplies (New York, 1898), and describe the filter beds at Hamburg, Germany.

29 (*b*) Consult Engineering News, August 23, 1894, and give a description of the filter beds at Mt. Vernon, N. Y.

30. Consult Zeitschrift für Hygiene, 1891, and give an account of an epidemic of typhoid fever at Berlin caused by a partial obstruction of filter beds by ice.

Chapter III.

WATER-SUPPLY SYSTEMS.

31. Classification.

Water-supply systems may be divided into two classes: gravity systems and pumping systems. A gravity system is one that collects the water of brooks in a reservoir and distributes it by gravity. A pumping system is one that elevates the water of a river or lake by means of pumps. The term "water-works" applies to both systems, and means a complete plant for the collection and distribution of a public supply.

The simplest gravity system has but one reservoir, which serves both to collect the water and to distribute it to the town. A more complete gravity system is that which has two kinds of reservoirs, one to collect and store the water, and the other for its distribution. The term storage system is also frequently used instead of gravity system for this class of water-works. Sometimes a town may have two or more distributing reservoirs for the supply of different sections, all being fed from the same storage reservoir. Sometimes a town may have two or more storage reservoirs which collect water from different brooks. In general, one collecting reservoir, with its distributing reservoirs and the pipes that lead from them, is called a gravity water-works.

Pumping systems are divided into two classes according to the method in which the water is distributed. The first class is where water is pumped to a reservoir from which it flows by gravity to the town; the second class is where the water is pumped directly into the main pipe leading to the town. In the first class opportunity is afforded for aeration and sedimentation, and the reservoir contains a supply for several days, so that the pump may be stopped when the river water is turbid. In the second class tanks and stand-pipes are sometimes provided, but these hold a small supply and there is little opportunity for sedimentation, since most of the water goes directly to the houses; accordingly, the water delivered by this method should be of very pure quality before it passes to the pumps.

The first public water supply in the United States was at Boston, Mass., where in 1652 a reservoir 12 feet square was constructed, to which water was brought through wooden pipes from neighboring springs. Nothing further appears to have been done until 1795, when the supply was increased and wooden pipes were laid for its distribution.

The second public water supply in the United States was at Bethlehem, Pa., where in 1754 a millwright named Hans Christopher Christiansen built a wooden pump which forced the water of a spring through a line of pitch-pine pipes 680 feet long to a wooden reservoir 70 feet above the spring, the pump being operated by an undershot wheel in a neighboring creek. In 1762 the wooden pump was replaced by three cast-iron pumps of 4 inches diameter and 18 inches stroke, gum-wood pipes used instead of pitch-pine, the height of lift increased to 112 feet, and from the reservoir pipes were laid to distribute the water to tanks and cisterns in the vicinity of the principal dwellings. The cost of this system, including the pump house, was £514 16s. 5d.

Prior to 1800 there were built in the United States of America only five public water-works, and prior to 1851 only 68. From 1851 to 1860 there were built 61, while from 1861 to 1870 the number was 104. After 1870 the number annually constructed increased with great rapidity, so that in 1880 there were 629 in operation. In the Manual of American Water-Works for 1888 there are described 1598 water-works in the United States, while the Manual for 1890-91 gives 2037. In the Manual for 1897 there are recorded 3196 water-works in the United States, which furnish both domestic supply and fire protection to 3480 towns, and also 462 furnishing a partial supply.

The sources of supply of these water-works embrace all the different kinds of surface and ground waters described in the last chapter. The relative proportions, as inferred from Flynn's analysis of the records for 1897, are about as follows: 40 per cent of the water-works use surface waters and 60 per cent ground waters; of the 40 per cent of surface waters about 6 per cent are from brooks and creeks, 27 per cent from rivers, and 7 per cent from lakes; of the 60 per cent of ground waters about 18 are from springs and 42 from wells.

The extent to which the different systems are used is also given roughly by the following figures for 1897: of gravity systems there are 25 per cent, and of pumping systems 75 per cent; the 75 per cent of pumping systems is divided into 15 per cent which pump to distributing reservoirs and 60 per cent which pump into the pipes either with or without tanks or stand-pipes. It is thus seen that the pumping systems are about three times as many as the gravity systems, and that the systems of direct pumping include over one-half the total number.

In the New England states, and also in the Pacific states, the number of gravity systems is about the same as that of the pumping systems. In the central and northwestern states, however, the number of pumping systems is twenty or thirty times as great as that of the gravity systems. In any particular case the local conditions determine the system to be used, that being selected which gives the best and purest supply at the minimum cost of construction and operation. The expense of construction is greatest for a gravity system, and the expense of operation is greatest for a pumping system.

32. Consumption of Water.

The amount of water used in a town depends mainly upon its population, but also upon the habits and occupations of the people. In designing a water-works it is customary to estimate the mean daily consumption per person and then to multiply this by the present or prospective population in order to find the probable mean amount that will be required. For this purpose the records of towns of similar character having water-works in operation for several years are to be consulted. As a rough estimate for approximate computations 100 gallons per person per day may be taken as a mean figure. The gallon used in this book is the American gallon of 231 cubic inches.

Ancient Rome had a daily supply of about 50 gallons per person. Modern European cities rarely exceed this amount. London uses 44 gallons per person per day, Paris 36 gallons, Berlin 30 gallons, while smaller cities like Geneva and Hanover use only about 25 gallons. Undoubtedly 50 gallons per person is an ample daily allowance, and the fact that the consumption of American cities is so much greater must be ascribed to waste rather than to reasonable use.

The mean daily consumption per person in Philadelphia was 68 gallons in 1880 and 132 gallons in 1890; in Chicago, 112 gallons in 1880 and 127 gallons in 1890; in St. Louis, 72 gallons in 1880 and 78 gallons in 1890; in Detroit, 130 gallons in 1880 and 155 gallons in 1890. In general the consumption of water shows a gradual increase in all the cities of the United States, but according to reliable estimates nearly one-half this is waste. This is demonstrated by the use of the meter plan, in which payment is made only for the actual amount of water drawn from the pipes; thus meters placed in a number of first-class apartment houses in Boston registered about 51 gallons per person per day, while those in moderate-class apartment houses showed 32 gallons, and in the lowest-class apartment houses only 17 gallons.

A small city generally uses less water per person than a large one; but the mean for American cities having 50 000 population is over 100 gallons per person per day. A manufacturing town has a high rate of consumption, and the same is true for a city with asphalt streets, since much water is used in washing them. In towns without factories and having a population of less than 10 000 the mean daily consumption will generally be between 60 and 100 gallons per person per day.

The daily consumption during July and August is from 15 to 20 per cent greater than the mean for the year, as in these months much water is used for sprinkling streets and lawns. In the northern part of the United States the daily consumption during January and February may be also 15 or 20 per cent higher than the mean, owing to the large amount that is wasted in order to prevent freezing of the pipes. If 100 gallons per person is the daily mean for the year the daily mean during these four months may be as high as 120 gallons per person.

On Mondays, when every household is at work on the weekly washing, the consumption may be put at from 20 to 40 per cent higher than the mean for the week. Accordingly, on the basis of 100 gallons per person as the daily mean for the year, the Monday consumption during very cold or very hot weather may be taken as from 140 to 170 gallons per person per day.

The amount required to extinguish fires is small when expressed as a daily mean, but an average fire requires about three hydrant streams, each delivering 200 gallons per minute. If two fires occur simultaneously in a town or fire district of 10 000 people the hourly consumption for fire purposes alone will be at the rate of 173 gallons per person per day. In general for each fire district the maximum available hourly supply for both fire purposes and domestic use should be at a rate from three to four times as great as that of the mean daily consumption.

The pressure under which the water is delivered in the streets is an important factor in all questions relating to fires. If the pressure be very high fire engines may not be needed, as the hose may be attached directly to the hydrants. If the pressure be very low it may not be possible to secure an effective fire service even with the use of engines, since a low pressure is always accompanied by a small discharge. In the houses a very high pressure greatly increases the waste of water, while of course a very low pressure furnishes an insufficient supply. Pressures over 100 pounds per square inch are high, and pressures less than 30 pounds per square inch are low.

33. Capacity of Storage Reservoirs.

When plans are to be made for a gravity supply there are two important preliminary questions to be discussed: first, what amount of water can be obtained; second, what storage capacity is needed for the supply of the town. The first question involves the preparation of maps of the neighboring watersheds, the collection of rainfall and run-off data, and certain reservoir estimates. The second question involves the considerations of consumption and pressure presented in the last article and more detailed estimates of storage capacity. The combination of the results of these two inquiries enables a decision to be made as to whether or not a given watershed will furnish a sufficient supply for the town. All the discussions of the last chapter regarding purity of the water should also receive careful attention during the progress of the inquiry.

As an example, suppose that a town of 6000 people, situated at A, on a stream too impure for domestic purposes, requires estimates to be made for obtaining a gravity supply by build-

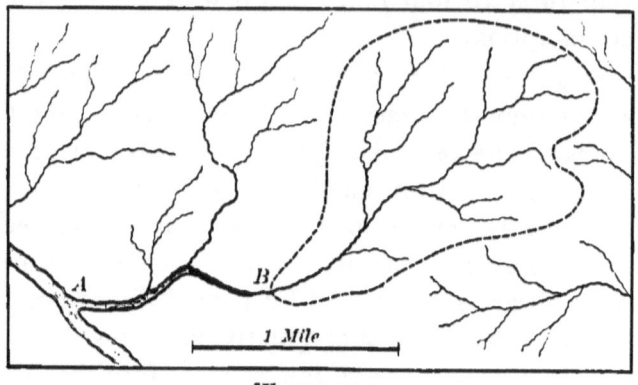

WATERSHED.

ing a reservoir at B to impound the run-off of a brook. From surveys and maps the area of the watershed above B is found to be 1390 acres. The mean annual rainfall is known to be 38 inches, of which about 45 per cent is run-off, the remainder going into evaporation and percolation, and hence the mean annual available storage is 647 000 000 gallons. The minimum annual rainfall, however, is 31 inches, and in such a year the available storage will be 528 000 000 gallons, or say a mean daily supply of about 1 400 000 gallons, which is more than 200 gallons per person. The flow of the brook at the driest season is found by measurement or estimation to be 150 000 gallons per day, or 25 gallons per person. It is accordingly clear that enough water can be obtained for the supply of the town if sufficient reservoir capacity be provided.

To estimate the capacity required, suppose that July is a wet month, August a dry month, and September a very dry month. Then during July the reservoir must store a quantity ample to supply the September demand. Let the mean daily consumption during August and September be 100 gallons per person, or a total of 600 000 gallons, and let the average daily run-off be 400 000 gallons in August and 150 000 in September. If the reservoir is full at the end of July it will not be full at the end of August, as the supply is 200 000 gallons less than the consumption; in September, moreover, the daily run-off received is 450 000 gallons less than the consumption. Accordingly, if the reservoir is to be half-full at the end of September it must have a capacity of 39 400 000 gallons.

The result of the preliminary inquiry in this case is that a sufficient supply for the driest season will be furnished by the watershed above B, provided that a reservoir holding about 40 000 000 gallons be constructed. To ascertain if such a reservoir is feasible a detailed survey of the site must be made and a map be drawn showing contours for every foot of vertical height. From this map the height and size of the necessary dams are determined, and then borings are made to ascertain the character of the foundations which these constructions require. Plans for the dam and its waste-weir, the pipe lines, distributing reservoirs, and street mains are also prepared, and finally an estimate of cost of the proposed gravity system is made. After the engineer has finished this work and made his report it remains for the town authorities to decide whether the money can be raised to carry out the execution of the project.

In this illustrative case the reservoir capacity required is about 80 times the mean daily consumption. This ratio is generally exceeded in the gravity systems of cities; thus in 1897 New York had a storage capacity of 38 000 000 000 gallons and a mean daily consumption of 230 000 000 gallons, giving a ratio of about 165; some other cities have a ratio higher than 200, while some small cities and towns run below 100 and occasionally below 50.

It is not easy to collect the data for an estimate of reservoir capacity, properly coordinate them, and draw correct conclusions; in fact it requires the training and good judgment of an experienced engineer to arrive at a sure decision. Other elements than those above outlined are also to receive attention, such as evaporation from the surface of the reservoir and the increase of supply due to the growth of the town. Estimates for several watersheds may have to be made and compared, and perhaps in the above case it might be cheaper to construct filter galleries or filter beds to purify the river water and distribute it by a pumping system than to carry out the proposed gravity supply. Anyone may make rough comparisons, but only the engineer can prepare such plans and estimates that a sure determination can be formed regarding the system which will furnish a pure and abundant supply and yet be the most economical in construction and maintenance.

34. Reservoir Dams of Earth.

Earthen reservoir dams have been built since the most ancient times and are still extensively used. When rock foundation is not at hand a masonry structure is impracticable, and an earthen embankment must necessarily be built. The figure shows a cross-section illustrating one of the best forms of construction. AB is the natural surface of the ground, and the trench CD is carried down several feet lower than this surface and filled with concrete or puddled clay, in order to prevent water from percolating under the dam. Above this trench the core is built of carefully selected material, rolled in layers concave upward, and on each side of this core common earth, usually called the frost covering, is placed. The width of the dam at the top is at least 15 feet, and the width at the bottom depends upon the height; if the height be 18 feet the bottom width should be 75 feet or more.

The core is the effective part of an earthen dam, and the material composing it should be such as to prevent the water from passing through it. For this purpose about 5 cart

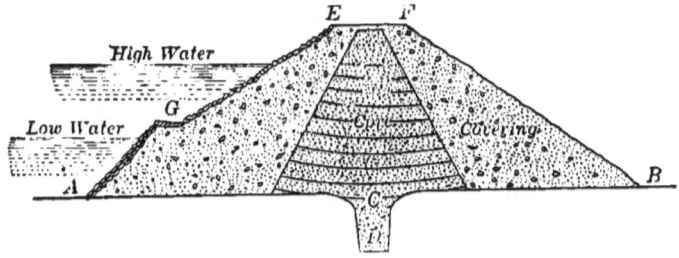

SECTION OF AN EARTHEN DAM.

loads of gravel are mixed with 2 loads of sand and 1 load of clay. The sand to a certain extent fills the spaces between the grains of gravel, while the smaller interstices are filled by the clay; thus when mixed and thoroughly rolled the 8 loads occupy a volume equal to about 6 original loads. The material must be kept well sprinkled during the rolling and the middle of each layer be kept somewhat lower than the sides. Instead of a core a thick wall of puddled clay is frequently used to prevent percolation.

The frost covering of earth is applied simultaneously with the construction of the core. This earth is any kind that may be at hand, as its office is not to prevent percolation of water, but merely to protect the core. The front slope BF is about $1\frac{1}{2}$ to 1, that is, $1\frac{1}{2}$ feet of horizontal projection to 1 foot of vertical projection, and it should be covered with grass. The back slope has a berm G just above low water, and this is wide enough so that a cart may run along it to make repairs. The slope AG is 1 to 1 or $1\frac{1}{2}$ to 1, and it is covered with riprap; the slope GE is about 2 to 1, and this is paved with thick stones, in order to prevent injury from the action of waves and ice.

If water runs over the top of an earthen dam its destruction surely follows, and hence a waste-weir or wasteway should be provided to carry off the excess of water; these will be discussed in Art. 36. Failures have also occurred by the percolation of water along the pipes which pass through the dams; to prevent this a good plan is to build a masonry culvert for carrying the pipes, the exterior surface of the culvert being rough and earth being puddled around it.

In 1874 occurred the failure of an earthen reservoir dam at Williamsburg, Mass., causing great damage to property in the valley below and the loss of 143 lives. The dam was about 550 feet long, 43 feet high at the middle, 16 feet wide at the top, and with side slopes of 1½ to 1. There was no core such as described above, but instead a rubble wall 2 feet thick at the top and 6 feet thick at the base was built through the middle. It was shown that this wall was of a rude character and imperfectly filled with mortar of a poor quality, and that its foundation was particularly defective, so that the percolation of water through and under it was the cause of the failure.

In 1889 one of the greatest disasters on record occurred by the failure of an earthen dam at the South Fork reservoir near Johnstown, Pa., 2142 lives being lost and property destroyed whose value was 3 500 000 dollars. The reservoir covered 407 acres and the watershed area was 48.6 square miles. The dam was about 18 feet wide at the top, with a slope of 1½ to 1 on the lower side and 2 to 1 on the upper side; the height was about 70 feet and the width at the base about 265 feet. It was shown that failure occurred not through any fault of construction of the main body of the dam, but entirely by reason of the insufficient size of the wasteway. This had been planned to be 150 feet in width and 10 feet in depth below the top of the dam, but as constructed its effective width was only 70 feet and, owing to the top of the dam having been subsequently lowered in height, its depth only 8 feet. During May 30 and 31 the rainfall on the watershed was between 6 and 8 inches, and for several hours the rate was greater than ⅜ inches per hour. The insufficient size of the waste-weir caused the water level in the reservoir to gradually rise until at 11:30 A.M. on May 31 it began to run over the top of the earthen dam. At 2:55 P.M. a portion of the dam, 400 feet wide at the top and 40 feet deep, broke away and a vertical wall of water 30 feet high swept down the narrow valley, destroying entire villages in its course. At 3:12 P.M. this flood of water reached the city of Johnstown, 12 miles down the stream and 250 feet lower than the reservoir, and there in a few minutes direful death and destruction were wrought.

It is an old saying that one failure teaches more than many successes. Certainly these two failures of earthen dams teach most emphatically two important lessons: first, that the construction of the dam and its foundations must be such as to prevent the percolation of water through or under it; second, that the width and depth of the waste-weir, or spillway, must be sufficient to discharge the accumulations of a very heavy rainfall. To secure these results plans alone are not enough, but constant and vigilant inspection of every phase of the work is required. As a consequence of these disasters boards of engineers have been instituted in some states to make annual examinations of reservoirs and recommend such improvements as they judge necessary for the public safety.

35. Reservoir Dams of Masonry.

Earthen dams far outnumber those of masonry, since the latter can only be constructed when a rock foundation is at hand. The rock must be entirely exposed, and be cut into trenches and steps, so that the dam may be thoroughly bonded with the rock, and thus all percolation of water be prevented.

The beds and joints of all the stones are to be entirely filled with hydraulic cement, and the bonding is to be such that no continuous horizontal joints are formed, in order that there may be no liability to sliding under the water pressure.

The shape of the cross-section of the dam will depend upon its height; when the height is less than 60 feet the trapezoidal form is commonly used. The back of the dam where the water pressure is applied may be vertical or have a slight batter, but the front has always a considerable batter. The thickness of the top of the dam ranges from 4 feet for low heights up to 15 feet or more for heights of 100 feet. The batter of the back and the thickness of the top being assumed, the thickness of the base is to be computed so that there may be ample security to resist the overturning action of the water.

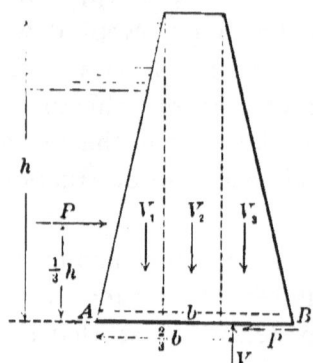

SECTION OF MASONRY DAM.

The figure shows the forces acting on a dam. The horizontal water pressure P is balanced by an equal resisting force P acting along the base AB. The weight of the dam, consisting of the weights of the parallelogram and two triangles, is $V_1 + V_2 + V_3$, and this is balanced by an equal upward resistance V. It is shown in treatises on construction that ample security will obtain when the thickness of the base is such that the distance from V to the toe B is one-third of the total thickness. Then, from the principle of mechanics that the sum of the moments of all the forces causing right-handed rotation around B must equal the sum of the moments of those causing left-handed rotation, this thickness b is computed.

The horizontal water pressure P for a dam one foot in length is shown by hydrostatics to be $31.25 h^2$ pounds if the dam be h feet in height. The point of application of this pressure is

$\frac{1}{3}h$ above the base of the dam. For example, let a dam be 48 feet high, then the horizontal pressure against one foot of its length is 72 000 pounds and the height of its point of application above the base is 16 feet.

Let a masonry dam 48 feet high and 1 foot long weigh 150 pounds per cubic foot and have a top thickness of 8 feet and a back batter of 1 inch to 1 foot. Then the base of the first triangle is 4 feet, its area 96 square feet, and the weight V_1 is 14 400 pounds. The area of the parallelogram is 384 square feet and the weight V_2 is 57 600 pounds. The base of the front triangle is $(b - 8 - 4)$ feet, its area is $24(b - 12)$ pounds, and the weight V_3 is $3600(b - 12)$ pounds. The total weight, which is equal to the upward reaction V, is hence $72\,000 + 3600(b - 12)$ pounds.

Taking B as a center of moments, the lever arm of P is 16, that of V_1 is $b - 2\frac{2}{3}$, that of V_2 is $b - 8$, that of V_3 is $\frac{2}{3}(b - 12)$, and that of V is $\frac{1}{2}b$. The forces P and V cause right-handed rotation around B, and the sum of their moments is

$$72\,000 \times 16 + [72\,000 + 3600(b - 12)] \times \tfrac{1}{2}b;$$

the forces V_1, V_2, and V_3 cause left-handed rotation around B, and the sum of their moments is

$$14\,400 \times (b - 2\tfrac{2}{3}) + 57\,600 \times (b - 8) + 3600(b - 12) \times \tfrac{2}{3}(b - 12).$$

Equating these two expressions and solving for b there is found 31 feet as the required thickness of the base. The batter of the front slope will therefore be $4\frac{3}{4}$ inches to each vertical foot.

The front of a masonry dam is frequently built with a less batter at the top than at the bottom, thus forming a broken or curved profile. For dams greater than 100 feet in height this is the general practice, as it is necessary in order that the stress on the stone near the toe B may not be too great. For very high dams the back is also curved in its lower portion. The design of these structures requires very elaborate computations, and these will be found explained in special treatises on engineering construction.

The principles governing concrete dams are identically the same as those for stone dams. The highest dam in America is the concrete structure at San Mateo, Cal., which impounds

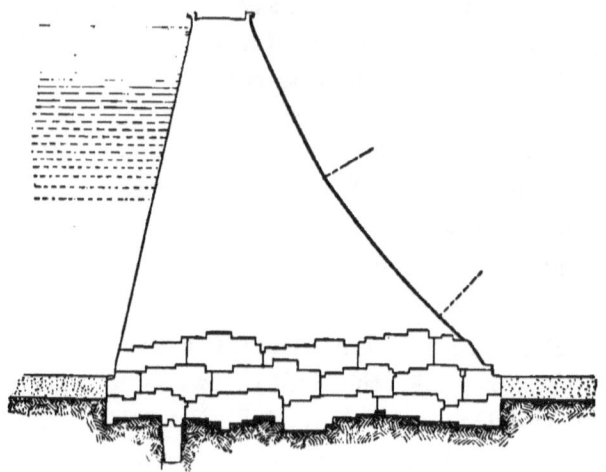

SAN MATEO CONCRETE DAM.

the reservoir for the water supply of San Francisco. Its height when completed is to be 170 feet and the top thickness 25 feet; the base thickness is 176 feet. The batter of the back is 3 inches to 1 foot, that of the front is about 5 inches per foot for a depth of 70 feet from the top, and then follows a curve of 258 feet radius, as shown in the figure. The length of the dam is 680 feet on top, and it is curved in plan so as to give an apparent increase in stability by virtue of the arch action. The concrete blocks near the base are very large, and they are made of a ⊤ shape, so as to thoroughly interlock.

There are but few masonry dams in the United States of a greater height than 50 feet, and most of these are built for irrigation reservoirs on the Pacific slope. At Reading, Pa., there is one 60 feet in height with a top thickness of 17 feet; and at Scranton, Pa., one 56 feet in height with a top thickness of 9½ feet. The structure proposed to be erected at Quaker Bridge on the watershed for the supply of the city of New York is designed with a height of 265 feet, a top thickness of 22 feet, and a base thickness of 216 feet.

36. Waste-weirs and Pipe Connections.

A waste-weir is an opening in the top of a dam to prevent the water from rising as high as the top. A masonry dam is often built without a waste-weir, for no damage is done by allowing the water to flow over the top if the bed of the valley for some distance below it be of rock, so that undermining of foundations cannot occur. For all earthen dams of storage reservoirs, however, a waste-weir, or wasteway, is absolutely necessary.

A waste-weir notch is in the top of a masonry construction built at one end of the dam. This masonry construction is in fact a masonry dam, usually of low height, the top being on the same level as the dam. Thus, in the figure, DD repre-

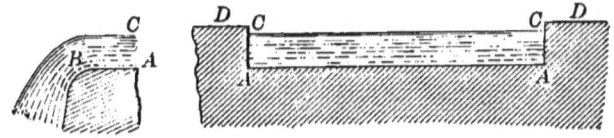

SECTIONS OF A MASONRY WASTE-WEIR.

sents the top, AA the sill of the waste-weir, and AC the depth of water flowing over the weir. The width of the dam on the sill is AB in the sectional view, where AC also shows the depth of the overflow. Let b be the width AA and H the depth AC; then, if b and H be in feet it is shown in hydraulic literature that the discharge in cubic feet per second is given by $q = 3.06bH^{\frac{3}{2}}$. From this formula the proper size of a waste-weir notch can be determined when the discharge q is known.

When a heavy rainfall occurs over the watershed the waste-weir should be able to discharge in one hour whatever part may reach the reservoir in that time. The longer such a rainfall continues the nearer should the discharging capacity of the waste-weir approach to the amount of rainfall on the watershed. For a large watershed a waste-weir designed to discharge one-third of an inch per hour on that watershed will generally give ample security, but for a small one, where the flood run-off occurs quickly, a somewhat higher figure should be taken.

As an example, the data of the watershed shown in the figure of Art. 34 may be used. The watershed area is 1390 acres, and supposing a rainfall of one inch per hour to occur and that six-tenths of this reaches the reservoir in one hour, the waste-weir must be sufficiently large to allow 840 cubic feet per second to pass over it. If the sill of the weir be 4 feet below the top of the dam, then $H = 4$ feet; and taking $q = 840$, the formula gives $b = 34.3$ feet as the width AA required. If H is to be 5 feet, however, then the width b is 24.8 feet. It is here seen that the determination of the data is the hardest part of the problem of designing the size of a waste-weir.

When rock is found at one end of the dam a canal may be cut through it, and this arrangement is called a wasteway. At the South Fork dam, whose failure was described in Art. 34, the wasteway was 176 feet long, the width at entrance 120 feet and at outfall 69 feet, while the horizontal bed was 8 feet below the top of the dam. The above formula may also be applied to this case, taking $H = 8$ feet and $b = 69$ feet, from which it is found that when the water was at the level of the top of the dam the discharge q was about 4780 cubic feet per second. The wasteway was, however, obstructed by fish screens, so that probably the actual discharge was only about 4000 cubic feet per second, while the amount entering the reservoir was about 8000 cubic feet per second.

The water pipes that run into a reservoir through an earthen dam should terminate in a gate chamber which is provided with valves for admitting and shutting off the flow. This gate chamber may be arranged so as to admit the water to it at different levels, for at certain seasons the water may be purer at one depth than at another. These openings are provided with screens to keep out fish. The gate chamber is necessarily of masonry, and it is hence generally built near one end of the dam where a good foundation may be obtained, and it often forms a part of the waste-weir structure.

An arched culvert extended through the embankment is one of the best ways for bringing the pipes to the gate chamber, a thick stop wall being built at the upper end to keep the water out of it. The exterior of the arch stones should be left rough and puddled clay be rammed around them, so as to prevent all percolation of water. This culvert carries not only the water main, but also a smaller pipe which may be used to draw off the reservoir when repairing or cleaning is to be done.

In all the masonry work of waste-weirs, gate chambers, and culverts hydraulic mortar of the best quality must be used and an efficient inspection be maintained to secure good material and workmanship. The strength of a structure is the strength of it weakest part, and hence if one part be defective through lack of proper inspection the security of the entire structure is correspondingly lowered.

37. AQUEDUCTS.

The water of a storage reservoir is carried to the distributing reservoirs by canals, aqueducts, or pipe lines. A canal is advantageous because it gives opportunity for the aeration of the water, but disadvantageous on account of the losses due to evaporation and percolation, as also on account of the liability to pollution; hence canals are rarely used. An aqueduct is an artificial channel constructed in masonry, but its top is covered; like a canal, it is laid out on a uniform slope, and unlike a pipe, it is never completely filled by the flowing water. The famous Roman aqueducts were carried across valleys on masonry arches in order to preserve a uniform slope; they were of rectangular cross-section, lined with concrete, and covered with either stone slabs or arched roofs. Aqueducts are now built only for the supply of large cities, a pipe line being sufficient to carry it in ordinary cases; they are built below the surface of the ground and are carried through rock in tunnels. The wood conduit is often used as synonymous with aqueduct, but it also applies to any large covered channel for carrying water.

The cross-sections used for modern aqueducts are generally of the circular and the basket-handle form. The circular section is used for small aqueducts less than about six feet in diameter. For larger diameters a basket-handle section with

 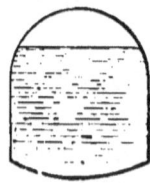 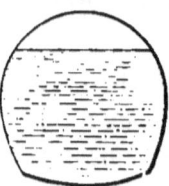

Cross-sections of Aqueducts.

vertical sides is used in rock or in soil so stiff as to exert little lateral pressure, while the sides are inclined inward in common earth; both of these have the bottoms formed of circular inverts of large radius.

Brick or stone is the material most generally used in aqueduct construction, and this is thoroughly bonded with hydraulic mortar. In all yielding soil there must be provided foundations of piles or concrete, in order to prevent settling. For the smaller circular sections cast iron or steel is used instead of brick. A masonry aqueduct is rarely filled with water up to the top, because it is not adapted to withstand a high outward water pressure, but an iron or steel one may be entirely filled if desired. In general, however, the word aqueduct implies that the water has a free level surface.

The quantity of water that an aqueduct will deliver depends upon the area of its cross-section, upon the surface in contact with water, and upon the longitudinal slope. Let a be the area of the cross-section, v the mean velocity of flow per second, and q the discharge per second; then $q = av$. Now let p be the length of the wetted perimeter, that is, the length of the inner part of the cross-section which is in contact with the water; then $a \div p$ is called the hydraulic radius of the cross-section and is designated by r. Let l be any length of the aqueduct in which the vertical fall is h; then $h \div l$ is called the slope and is designated by s. The formula for the velocity v may now be stated as $v = c\sqrt{rs}$, in which c is a number which depends upon the roughness of the surface and other circumstances. If the surface were perfectly smooth the mean velocity v would continually increase, but owing to the friction caused by the roughness it remains constant unless r and s change in value.

The value of the coefficient c is to be found by consulting the discussions of recorded gagings given in treatises on hydraulics. If the aqueduct has a cement lining, and r be taken in feet and v in feet per second, the following values of

COEFFICIENTS FOR AQUEDUCTS.

Hydraulic Radius in Feet.	$s = 0.00005$	$s = 0.0001$	$s = 0.0002$	$s = 0.0004$	$s = 0.001$
$r = 1$	114	120	123	125	127
$r = 1.5$	126	130	133	135	136
$r = 2$	135	138	140	141	142
$r = 3$	147	148	149	149	150

c may be employed. For example, if an aqueduct be laid on a slope of one foot per mile the value of s is $1/5280 =$

0.0001894; if its cross-section be 48 square feet and its inner perimeter 25 feet the value of r when running full is 1.92 feet; then from the table the coefficient c has the value 139.

The capacity of an aqueduct is its maximum discharge, and this occurs when it is nearly but not quite full. For the case where the slope s is 0.0001894, the hydraulic radius $r = 1.92$ feet, and $c = 139$ the mean velocity v is found by computation to be 2.65 feet per second. Then the discharge q is 127.2 cubic feet per second, or 82 200 000 gallons per day; this capacity is sufficient for the supply of a city of 500 000 people. If the slope of this aqueduct be 4 feet per mile its mean velocity and discharge will be double the above figures; a slope as great as this is, however, very uncommon, although greater slopes are said to have been used in the old Roman aqueducts.

The old Croton aqueduct for the supply of New York, built in 1842, has a cross-section of 53.3 square feet, a length of 38.1 miles, an ordinary slope of 1.11 feet per mile, and a capacity of 98 000 000 gallons per day; it enters the city on the beautiful masonry structure across the Harlem River known as High Bridge. The new Croton aqueduct, completed in 1890, has a cross-section of about 160 square feet and a capacity of about 300 000 000 gallons per day; it enters the city by a tunnel or inverted siphon under the Harlem River. Baltimore, Boston, and a few other large cities also have aqueducts.

It is much cheaper to build an aqueduct than several pipe lines of the same capacity; for example, it requires five or six pipes of 4 feet diameter to carry the same amount of water as one aqueduct 8 feet in diameter, and the cost of five pipe lines would far exceed that of the aqueduct. The general rule for circular cross-sections of the same degree of smoothness is that the capacity varies as the square root of the fifth power of the diameter; thus, if there be two sections of 2 feet and 8 feet diameter the larger one carries 32 times as much water as the smaller one.

38. PIPE LINES.

An iron or steel pipe is adapted to stand the outward pressure of water, and hence may be carried down and across a valley, following the undulations of the surface, whereas an aqueduct must wind around so as to keep the slope nearly

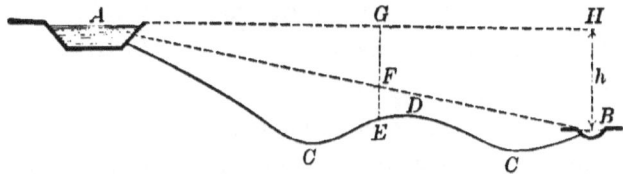

PIPE LINE FROM STORAGE TO DISTRIBUTING RESERVOIR.

uniform. Thus, if A be a storage and B a distributing reservoir the pipe line $ACDCB$ is laid on the shortest practicable route. The ancient Romans were not ignorant of the fact that this could be done, as some suppose, but they had no iron or steel pipes, and hence were compelled to use the aqueduct method.

Cast-iron pipes are the most common and may be obtained of any diameter up to 5 feet; each pipe is 12 feet long, one end having a bell-like flange into which the end of the next pipe is placed and rendered water-tight by a lead joint. Steel pipes are made in sections, each section being formed of a single plate with a longitudinal riveted joint and the different sections being riveted together by transverse joints. Small steel pipes have long been used in California in mining operations, and in 1876 one of 36 inches diameter was laid at Rochester, N. Y.; since 1892 several lines of larger diameter have been constructed, notably the East Jersey system of 36, 42, and 48 inches diameter, and a line at Allegheny, Pa., of 60 inches diameter. On the Pacific slope wooden pipes are also used; these are made of redwood staves, which are bound with adjustable steel hoops.

On the pipe line AB there should be a valve both at A and at B, in order that the flow may be regulated or be entirely shut off. At a low point like C a mud valve should be placed, in order to clean out deposits that may be formed. At a high point like D an air valve is provided, in order to allow the escape of the air which collects there. The line AB, drawn on a uniform slope from the water level in A to that in B, is called the hydraulic grade line. No point of the pipe should be above this hydraulic grade line, for if so a retardation in flow is liable to occur.

When the valve at A is open and that at B is closed the pressure of the water at every point in the pipe is due to the head of the water level in A; thus at E the pressure is due to the head EG and at B the pressure is due to the head BH. When the valves at both A and B are fully open the condition of things is very different, and the pressure at any point in the pipe is due to the head measured up to the hydraulic grade line; thus at E the pressure is due to the head EF and at B the pressure is that due to the depth of water vertically above the end of the pipe.

The method of the last article may, with proper modifications, be used for computing the flow in pipe lines. The length l is here the length of the pipe measured along it from A to B, and the fall h is the difference in level between the water surfaces at A and B or the head BH; then the slope s is h/l. If the diameter of the pipe is d the area of its cross-section is $\frac{1}{4}\pi d^2$ and its circumference is πd; hence its hydraulic radius r is $\frac{1}{4}d$. The mean velocity of flow may then be written $v = c\sqrt{dh/4l}$, and after this has been computed the discharge per second, q, is found from $q = \frac{1}{4}\pi d^2 \cdot v$. The coefficient c will, as before, vary with the roughness of the surface, the size of the pipe, and the slope of the hydraulic grade line.

For a new cast-iron pipe, coated with tar or asphalt, the following values of the coefficient c may be used; these, like those of the last article, are derived from the discussions of Kutter. For an old pipe the values of c, owing to incrustations and deposits, may be 20 or 30 per cent less than the

COEFFICIENTS FOR CAST-IRON PIPES.

Diameter in feet.	$s = 0.0001$	$s = 0.0002$	$s = 0.0004$	$s = 0.001$	$s = 0.01$
$d = 0.5$	73	79	83	87	88
$d = 1$	92	98	102	104	106
$d = 2$	112	117	120	122	124
$d = 3$	124	129	131	133	134
$d = 4$	132	136	138	139	141
$d = 5$	140	141	142	143	144

tabular values. For a steel pipe, owing to the frictional resistance of the rivets, the values of c, and hence also the velocity and discharge, are also less than those with smooth inner surfaces.

As an example, let it be required to compute the discharge through a new clean pipe 30 inches in diameter and 16 400 feet long under a head of 49.2 feet. Here $d = 2.5$ feet, $h = 49.2$ feet, $l = 16\ 400$ feet, $s = 49.2/16\ 400 = 0.003$, and hence from the table $c = 128$. Then by computation the mean velocity v is found to be 5.54 feet per second, and the discharge q to be 27.2 cubic feet per second, which is equivalent to about 17 500 000 gallons per day.

The inverse problem, to find the diameter of a pipe to carry a given discharge, is also solved by the use of the above formulas and table. For example, let it be required to find the diameter to carry 8 500 000 gallons per day when the length of the pipe is 25 400 feet and the head 127 feet. Here the slope s is 0.005, and the value of c may be assumed as 100; the discharge q is 13.15 cubic feet per second. By transformation the above formula for q becomes $d = (8q/\pi cs^1)^{\frac{1}{5}}$, from which the diameter d is found to be 0.89 feet; this result compared with the table shows that the assumed value of c is not far from correct and that a diameter of one foot will give a somewhat greater discharge than that required. To allow for all contingencies it is hence well to use a 12-inch pipe.

While the above gives a brief outline of the method of designing the diameter of a pipe line, it should be said that the selection of the proper coefficient is a matter that involves much judgment and experience. The same is true of the location of the route, and even of the laying of the pipe in the trenches. The thickness of the pipe will be different under different heads, and this depends also upon the diameter and the kind of material; in Arts. 44 and 45 the question of thickness is further discussed. All these points require careful consideration by the engineer, in order that the pipe line may have full capacity, an ample factor of safety, and the highest degree of economy.

39. Distributing Reservoirs.

A distributing reservoir is small compared with a storage reservoir, as the latter is required to hold sufficient water for two or three months' consumption, while the former is to provide only for a few days. Thus, in 1897 the city of New York had 13 storage reservoirs with a capacity of 38 000 000 000 gallons and 5 distributing reservoirs with a capacity of 1 350 000 000 gallons. In times of heavy rainfall it is advisable to shut off the connection with the storage reservoir, in order to allow time for sedimentation, and thus in general it is well that the distributing reservoirs should hold a supply sufficient for nearly a week. A large city has several distributing reservoirs for the supply of different districts, while a small town has perhaps but one. Two reservoirs are preferable to one, however, as one of them may be thrown out of use while the other is undergoing repairs or cleaning. A connection of the town with the storage reservoir is also always advisable, so that in an emergency water may be drawn directly from it.

The distributing reservoirs of a pumping system differ in no essential respect from those of a storage system, except that they should usually have a higher capacity, in order to allow time for sedimentation. Here one reservoir is decidedly inadvisable, but there should be at least two, one of which receives the water from the pump while the other distributes it to the town. The connection between the two reservoirs may then be shut off for a day or two, either when repairs are necessary or when the water pumped is turbid owing to river floods. Three distributing reservoirs form a better arrangement still. For example, at South Bethlehem, Pa., the water is pumped from the Lehigh River to a reservoir of 12 000 000 gallons capacity, from which it passes to two smaller ones having a combined capacity of 3 000 000 gallons; here the large reservoir acts like the storage reservoir of a gravity system, and pumping may be discontinued for several days when the river water is turbid, while at the same time either one of the smaller reservoirs may be thrown out of use whenever necessary.

A distributing reservoir is built upon a hill which has sufficient elevation to give the required pressure throughout the town. A storage reservoir made by damming a valley requires no excavation further than that of removing the vegetable matter in the top layer of the soil, but a distributing reservoir on a hill involves considerable excavation. The most economical location is generally near the brow of the hill, where the work will be partly excavation and partly embankment, as shown in the figure. Extraordinary care must be taken that this embankment may have no tendency to slide and that no water may percolate through it. The shape of the reservoir will depend upon the configuration of the ground and the capacity required. The depth of water in the middle usually ranges from 15 to 30 feet.

Most distributing reservoirs require to be lined with concrete, in order to prevent leakage. The earth over the entire inner surface should be thoroughly rolled and then a layer of about one foot of clay is applied and consolidated by tamp-

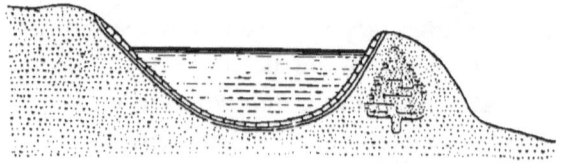

SECTION OF A DISTRIBUTING RESERVOIR.

ing. Over this a layer of concrete is placed, the thickness of this being also about one foot. Sometimes an asphalt lining is added, since concrete is more or less pervious to water under a head of 20 or 30 feet.

When a distributing reservoir is filled from an aqueduct the regulation of the entering water must be done entirely at the storage reservoir. When it is filled from a pipe line there should be a valve at both reservoirs, and the closing of one will shut off the supply. If both valves are open the maximum discharge of the pipe occurs, but the energy of the flow is scarcely sufficient to raise the water a foot above the outlet end. In order to throw the entering water into the air, so that it may absorb oxygen, the discharge must be diminished by putting nozzles on the outlet end; this increases the height of the hydraulic grade line throughout and gives a pressure head which will cause fountain streams to rise.

The gate chamber at the distributing reservoir is arranged in a manner similar to that at the storage reservoir. A waste pipe to draw off the water is laid from the lowest part of the bottom to a convenient point of discharge beyond the embankment. All distributing reservoirs should be cleaned of mud deposits once a year, and some require cleaning at shorter intervals. It has occasionally happened that a reservoir has become polluted with sewage; in such a case its surface must be most thoroughly washed with bromine water, in order to kill all the bacteria.

When filter beds or mechanical filters are used in a gravity system these are generally placed between two reservoirs, one of which receives the water from the storage reservoir and delivers it to the filter plant while the other receives the purified water and distributes it to the town. In a pumping system with distributing reservoirs a similar arrangement may be followed, or the filter plant may be near the river and the purified water be pumped to the reservoir. When the system of pumping to a tank or stand-pipe is employed it is necessary that the filtering should be done before the water passes through the pumps.

40. Pumps and Pumping.

For raising water out of an open or driven well a suction pump may be used if the lift is not over 30 feet. For a higher lift the combined suction and force pump must be used, and this must be placed within 30 feet of the water level. If a perfect vacuum could be formed it would be possible to raise water 34 feet by suction, but in practice, on account of leakage in valves, this limit cannot be attained. Through the lift above the pump the water is forced up by the pressure exerted by the piston; if this exerts a pressure of p pounds per square inch the theoretic height of lift in feet is $2.304p$, but on account of frictional resistances this also cannot be attained.

A common type of single-acting pump is that which has a horizontal cylinder and a solid piston. When the piston moves through the stroke from C to D the upper valve closes and the lower one opens; thus a partial vacuum is formed and the atmospheric pressure causes the water to rise from A and fill the cylinder. When the piston moves through the stroke from D to C the lower valve closes and the upper one opens and the water is forced up through the discharge pipe to the outlet at B. The effective work done by the pump is that of raising the water through the distance AB, and the total work

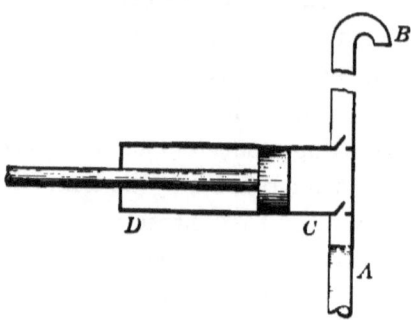

SINGLE-ACTING PUMP.

is this effective work plus that required to overcome the frictional and retarding resistances.

Two single-acting cylinders placed side by side and connected with the same suction pipe and discharge pipe are generally called a duplex pump. The pistons move in opposite directions, so that when one is forcing water through the discharge pipe the other is drawing water up the vacuum pipe. Three cylinders may be also used, in which case it is called a triplex pump. In order to cause a uniform flow in the discharge pipe, as also to render the operation more smooth, an air vessel is generally provided; into this the flow from the pump cylinders passes and out of it the discharge pipe rises. The air in the vessel, being compressed by the water pressure, acts like a cushion to absorb shocks and the flow up the discharge pipe becomes perfectly continuous.

A double-acting pump is one that forces the water in both strokes of the piston. When the piston moves from C to D the valves E and F open, while e and f close, and the flow of water is in the direction of the letters $AEFB$. When the piston moves from D to C the valves e and f open, while E and F close, and the water follows the direction $AefB$. Here, as in all cases, the effective work of the pump is that of rais-

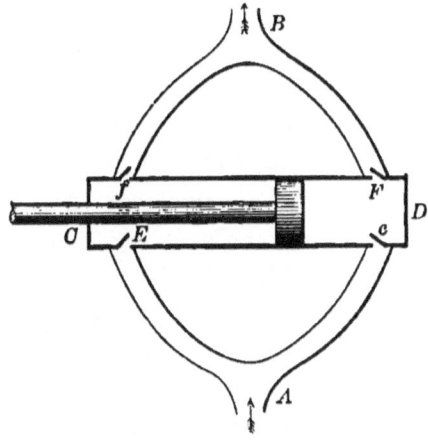

DOUBLE-ACTING PUMP.

ing the water from the level below A to the desired elevation above B.

The capacity of a pump is measured by the greatest amount of water it can deliver per day; thus a pump of 3 000 000 gallons capacity is understood to be one that can raise 3 000 000 gallons of water in 24 hours. This gives, however, little idea of the work done by the pump unless the height of lift or the pressure that it maintains is also stated. A pump of 3 000 000 gallons capacity lifting water through a height of 100 feet becomes merely one of 1 500 000 gallons capacity if the height of lift is 200 feet.

The power of a pump is the number of horse-powers it can deliver, and one horse-power is the performance of 550 foot-pounds of work in one second. Thus, if 1 200 000 gallons of water is to be raised through a height of 230 feet in 24 hours the weight to be lifted in one second is $116\frac{1}{4}$ pounds and the power required is 48.6 horse-powers. The effective power of the pump must, however, be considerably higher than this, as work is lost in overcoming the frictional resistances due to the flow in the pipe as well as in the pump cylinders.

Pumps are sometimes driven by water or electric power, but steam power is used for those of high capacity. The term pumping engine includes not only the water cylinders, but also the steam cylinders, all the steam apparatus being in fact regarded as a part of the pumping plant. These pumping engines will be discussed in the next article.

The air lift pump, introduced about 1890, is an apparatus for increasing the flow of deep wells by forcing compressed air through a tube to a considerable distance below the ordinary level. The tube is carried down inside a vertical pipe and the air issues from the end of the tube and forms bubbles in the water. The entire column of water in the pipe, being filled with bubbles, is lighter than a column of common water, and hence rises to a greater height. In this manner water whose natural level is 50 feet or more below the surface of the ground may be caused to rise above that surface. The cost of pumping by this method appears to be rather high, and it is unlikely that it will supplant the common pumps which are operated by steam or electric power and sometimes also by gas or air engines. The air lift system, however, thoroughly aerates the water, and hence in some cases improves its quality.

41. Pumping Engines.

A pumping engine is one run by steam, the pumps, steam cylinders, and boilers forming one plant which must be considered as a whole in discussing its economy, since the coal burned under the boilers is a large item in the operating expenses. Such a plant should not only have a horse-power sufficient to deliver the required quantity of water under the given pressure, but the cost of installation and operation should be such as to be most economical. A cheap pumping engine will consume more coal than an expensive one, just as a cheap coat will require more repairs than one of higher price. The amount of coal consumed annually is therefore as important an item to be considered as that of the first cost of the pumping plant.

The duty of a pumping engine is the number of foot-pounds of work that it can do with an expenditure of 100 pounds of coal. This is the old definition of duty, as first stated by Watt, and it is still very useful in general discussions. Thus, when it is said that a pumping engine has a duty of 120 000 000 it is meant that 120 000 000 foot-pounds of work can be performed by it by burning 100 pounds of coal under the boilers. The ambiguity in this definition regarding the meaning of the word coal has, however, led to a more precise definition, but the old definition will agree with the modern one if the coal be understood to be of such quality that one pound of it is capable of generating 10 000 British heat units, or 7 780 000 foot-pounds of work.

The steam engine is a very wasteful utilizer of the energy in coal. Nearly one-half of this energy goes up the chimney, and about one-third is lost in the exhaust steam, so that more than one-fifth of it is rarely utilized. A pumping engine of 150 000 000 duty utilizes 19.3 per cent, one of 100 000 000 duty utilizes 12.9 per cent, and one of 50 000 000 duty utilizes only 6.4 per cent. The higher the duty the greater is the amount of work that can be done with one pound of coal. Using coal of the standard quality, as above defined, 100 pounds burned in one hour produces 75.8 horse-powers with a 150 000 000-duty engine, 50.5 horse-powers with a 100 000 000-duty engine, and 25.2 horse-powers with a 50 000 000-duty engine. Or, if 1000 pounds of coal are required to do a given amount of work with a 150 000 000-duty engine 1500 pounds will be required with a 100 000 000-duty engine and 3000 pounds with a 50 000 000-duty engine.

A high-duty engine is an expensive one and a low-duty engine is a cheap one. In any particular case the question of consumption will determine the required capacity, and this together with the pressure will determine the horse-power. It is then necessary to go into the market and ascertain the prices of pumping engines of different duties to do the work required. The cost and calorific power of the coal to be used are also to be ascertained. Then the prices of installation are discussed in connection with the costs of operation, and thus the most economical engine may be selected. If coal is very dear a high-duty engine is best; if coal is very cheap an engine of low duty will be preferable.

For example, take a town whose mean daily consumption of water is 2 500 000 gallons per day, which is to be delivered into the pipes under a pressure of 50 pounds per square inch at the pump, while the suction height is 5 feet. Here 50 pounds per square inch is equivalent to a head of 115 feet, so that the work required is the same as if the supply were lifted through a height of 120 feet. The capacity of this pumping engine should be at least 5 000 000 gallons per day in order to meet the demands of the Monday consumption, and accordingly the maximum work to be done is 5 013 000 000 foot-pounds per day, and the effective power required is closely 106 horse-powers. On obtaining prices or bids for

COMPARISON OF ECONOMY OF PUMPING ENGINES.

Duty. Millions of foot-pounds.	Cost of Pumping Engine.	Annual interest and sinking fund at 7 per cent.	Annual cost of coal at $3 per ton.	Annual payment for interest and coal.
150	$24 000	$1680	$817	$2497
140	20 000	1400	876	2276
130	17 000	1190	942	2132
120	15 000	1050	1022	2072
110	14 000	980	1114	2094
100	13 000	910	1225	2135
90	12 000	840	1362	2202

pumping engines of this capacity and power it is found that the cost of those of different duties are as given in the second column of the table. If the annual interest on this cost, together with a contribution to a sinking fund in order to repay the principal after a certain number of years, be 7 per cent the annual expense for this purpose is given in the third column. Now as the mean consumption is 2 500 000 gallons per day under a head of 120 feet, the mean work per day is 2 506 000 000 foot-pounds, and 100 times this, divided by the duty, gives the number of pounds of coal used per day; whence, for coal at $3 per gross ton, the annual cost of coal is found and recorded in the fourth column. The sum of the annual interest charge and annual cost of coal gives the last column of the table, from which it is concluded that the 120 000 000-duty pumping engine, whose cost is $15 000, is the most economical.

It is here very clearly seen, as in so many other cases in the preceding articles, that expert knowledge is necessary in order to purchase a pumping engine and secure the highest degree of economy. Of course anyone can buy a pump, but only the experienced engineer is able to buy one which will be just suited to do the required work, and which at the same time will give the highest degree of economy. To insure that the pumping engine installed conforms to the specifications under which it is bought, a test must be made before paying for it. This test will determine the calorific capacity of the coal, the indicated horse-power of the steam cylinders, the capacity of the pump working under the assigned pressure, the efficiency of the steam and water machinery, and the duty of the combined plant. If the results of such a test show that the apparatus does not meet the requirements of the specifications it is clear that a claim for reduction in the stipulated price should be made, or that such alterations should be demanded as will cause the plant to fulfil those requirements.

42. Pumping to Reservoirs.

Pumping systems of water supply are of two classes, one having a distributing reservoir where sedimentation may occur, and the other having none. The first class will here be described. As far as the distributing reservoir is concerned the remarks in Art. 39 explain the general method of construction. The water from the pump enters the reservoir by one pipe, while other pipes distribute it either to other reservoirs or to the town. Thus in the following figure CB represents the pipe leading from the pump C to the reservoir B, while BD is the pipe which carries the water out of the reservoir.

A river is the most common source of supply, and a well should be excavated in or near its bank for the reception of the suction pipe. Such a well is usually lined with rubble

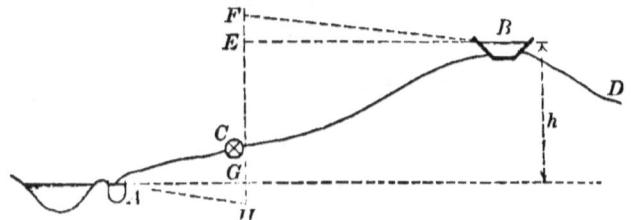

PUMPING FROM RIVER TO RESERVOIR.

masonry without cement, so that the water may enter through the walls as well as through the bottom, and it is covered so that the refuse of floods may not enter at the top. In some cases the well is built below the river bottom instead of on the bank, and in other cases it is built on the river bottom and the water admitted through iron gratings.

The pump C must be located so that the vertical height of the pump cylinders above the water in the well is less than 30 feet and preferably less than 20 feet. The vertical height of the reservoir B above the pump is not often greater than 300 feet and usually less than 250 feet. The location of the reservoir with respect to the pump and the town depends, of course, on the topography of the region, and this will determine the lengths of the force and distribution pipes.

The power of the pump must be sufficient to lift the water through the height h, which is the same as GE, or the difference in level between the water surfaces in the well and reservoir, and also to overcome the frictional resistances. Of these resistances the most important is that of the friction of the water in the pipe; this increases with the square of the velocity of flow, and hence may be made small by using a pipe of large diameter. In the figure the line CE represents the pressure-head at C when there is no motion of water in the pipe CB, and CF represents the pressure-head when the pump is at work; thus EF is the head which is required to overcome the friction in the force pipe. In like manner the apparent height of lift for the suction pipe is CG, but on account of friction this is increased to CH. Thus the pump must have an effective power sufficient to overcome the total head between H and F, as well as the resistances due to its pistons and valves.

It is shown in treatises on hydraulics that the friction-head caused by the mean velocity v in a pipe of length l and diameter d depends upon the roughness of the inner surface of the pipe, its diameter, and the mean velocity of flow. The following table gives values of this friction-head for 100 feet

FRICTION-HEAD FOR 100 FEET OF PIPE.

Diameter in feet.	$v = 1$	$v = 2$	$v = 3$	$v = 4$	$v = 6$	$v = 10$
0.25	0.20	0.70	1.46	2.40	5.37	
0.5	0.09	0.32	0.70	1.14	2.46	6.22
0.75	0.05	0.21	0.45	0.73	1.57	3.94
1	0.04	0.15	0.32	0.55	1.12	2.80
1.5	0.02	0.09	0.20	0.33	0.67	1.66
2		0.06	0.13	0.21	0.45	1.09

in length of clean cast-iron pipe, and from it the friction-head for any other length is readily obtained. The values of v at the tops of the columns are in feet per second. For example, if a pipe 12 inches in diameter and 12 570 feet long discharges 52 850 gallons per hour the mean velocity in feet per second is $v = 2.5$ and the table gives 0.235 feet as the friction-head for 100 feet of pipe; whence the friction-head for the given pipe is $125.7 \times 0.235 = 29.5$ feet. If the height of lift is 120 feet the pump must hence be able to overcome the pressure due to a head of 149.5 feet.

As an illustration of the effect of the size of the pipe on the horse-power of the pump suppose that 500 000 gallons per hour is required to be raised through a vertical height of 230 feet from the well A to the reservoir B, the total length of the pipe being 4200 feet. Here the discharge per second through the pipe is 1.86 cubic feet, or closely 116 pounds, and the power required to lift this through 230 feet is 48.2 horse-powers. If the pipe be 6 inches in diameter the mean velocity is 9.47 feet per second and, from the table, the friction-head is about 233 feet, or more than the given lift, so that the pump would have to exert about 97 horse-powers. If the pipe be 12 inches in diameter the mean velocity is 2.37 feet per second, the table gives 8.9 feet as the friction-head, and 50.2 horse-power is required. The great advantage of using a large pipe is here plainly shown, and in any given case such a size is to be selected as will render a minimum the total annual expenses, both for interest and operation, of the entire pumping system.

43. Direct Pumping.

The second class of pumping systems is that where the pump delivers the water directly into the main which supplies the town. This class has three subdivisions: first, where a part of the water goes to a tank which holds a supply sufficient to allow the pump to stop for a day or several hours; second, where a stand pipe is placed near the pump which holds a supply sufficient to allow a stoppage of an hour or two; third, where no tank or stand pipe is used, so that the supply of the town is entirely dependent upon the pressure exerted by the pump pistons.

The method of pumping to a tank is illustrated in the following figure, where A is the pump well, C the pump, and

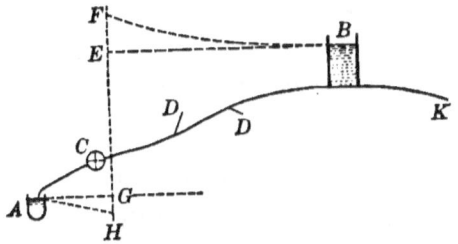

PUMPING WITH TANK.

CB the pipe running to the tank and through the town, where lateral pipes, marked D, carry the water into different streets. If there be no water taken from the pipe between C and B, and a distribution pipe K delivers it out of the tank B, the system is the same as that described in the last article. In many cases the water is delivered both out of the main pipe and out of the tank, thus making a mixed system. As far as pumping capacity and power are concerned there is little difference between the two classes. In each GE is the height of lift and HF is the total effective head that the pump must overcome.

The method of pumping with a stand pipe, illustrated in the next figure, consists in forcing the water into the stand pipe PE until it is nearly filled. The head PE then produces the necessary pressure to cause the distribution through the mains. The best location for the stand pipe is near the pump, as it then acts somewhat like an air vessel to neutralize the effect of shocks, the water rising in it when the flow is suddenly checked instead of bringing a direct shock on the pump cylinders. Here, as before, EF represents the head which measures the frictional resistances in the street mains beyond P. If the pump stops the water level at E immedi-

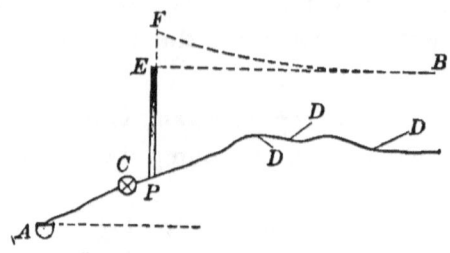

PUMPING WITH STAND PIPE.

ately falls with the consumption, and after a short time the pressure in the street mains is materially lowered.

The method of direct pumping without the use of tank or stand pipe may be illustrated by omitting the tank in the first of the above figures; here the pump must maintain a pressure corresponding to the head CF, of which EF is expended in overcoming the friction in the mains. If the stand pipe in the second figure be omitted it also represents the method. The regulation of the pumping engine in this method is effected through the water pressure itself by means of a weighted vertical piston or some other suitable device. When no water is being drawn in the town the pump is motionless; when a little is drawn it moves slowly to keep up the supply and pressure; when the draft is heavy it moves rapidly. Whatever be the consumption the regulation should be such as to maintain in the pipes a pressure equivalent to the head below the horizontal line BE. It is seen that duplicate pumping engines are almost indispensable, for if there be only one a stoppage for repairs deprives the entire town of its supply.

The computation of the power of the pumping engine required to furnish the maximum hourly consumption is the same for these three methods and differs from that of the last article only in regard to the friction-head, which is generally not so great, since the velocity of the water in the mains decreases with the distance from the pump. The head CE will be determined by the pressure which is to be maintained; if 100 pounds per square inch is required, then this head is 230 feet. The given consumption and the size of the pipes at C or P determine the mean velocity at that point, and from this, by the help of the table of the last article, a value of the friction-head is found. One-half or one-third of this will, in general, be all that is really expended in days of ordinary draft, but if a fire occurs the flow may be so concentrated in certain mains that the full value of the computed friction-head is reached. It is, therefore, probably best to take it as given by the table in cases of design, but in the investigation of an existing system a detailed investigation will be necessary to determine an accurate value. If the street mains are large the head lost in friction will be small, but if mains of small diameter be used the friction may become so great as to require an engine of high power whose coal consumption may be a serious item of expense.

The method of direct pumping with or without a stand pipe is adapted to towns on level ground, and is extensively used in the prairie regions of the central states. Chicago in 1897 had about 30 pumping engines at seven different stations which delivered the water of Lake Michigan throughout the city, the total capacity of the pumps being 358 000 000 gallons per day; one of these engines pumped to a reservoir of small capacity, and at two stations there were stand pipes 138 and 167 feet in height; thus the greater part of the service is by the third method of direct delivery. Where the topography admits, the method of pumping to tanks or that of pumping to distributing reservoirs is generally preferred on account of the advantage of maintaining even a small amount of storage for cases of emergency. There is, of course, nothing in any system which renders it universally more advantageous or economical than another, but the engineer, in each particular case, selects that system or combination of systems which the local conditions will render the most reliable and at the same time give the lowest annual outlay for the interest on construction and the expenses of operation.

44. Tanks and Stand Pipes.

A tank may be built of masonry, the cross-section being rectangular and the side walls of sufficient thickness to safely withstand the lateral pressure of the water. Such a construction is, however, too expensive when the depth of water is over 10 or 15 feet, on account of the very thick walls that are required. For greater heights the common method is to build a structure of circular cross-section with wrought-iron or steel plates riveted together like a large steel water pipe. It is connected to the masonry foundation by steel angles or knees, and all the joints and connections must of course be water-tight. The top is usually left open, but it is best to put a roof over it in order to prevent the multiplication of algæ under the action of sunlight, and in very cold regions the tank should be entirely housed with wood or brick in order to decrease the liability to freezing.

The height of a tank depends upon the pressure that is to be maintained in the town and the elevation of the hill on which it stands, and its diameter depends upon the quantity of storage that is required. Tanks larger than 50 feet in diameter and 60 feet high are not common; one of this size will hold about 880 000 gallons, and if all of this is available for consumption it is sufficient for one day's supply for a town of 9000 people. A tank of smaller diameter may be built higher than 60 feet, since the thickness of the steel plates decreases with the diameter; thus a tank might be 30 feet in diameter and 100 feet high, and its capacity would be about 490 000 gallons.

44. TANKS AND STAND PIPES. 127

A stand pipe may be built of steel plates exactly like a tank, but its diameter is less and its height usually greater. The tallest stand pipes are about 250 feet in height, but the number higher than 200 feet is small. As the function of a

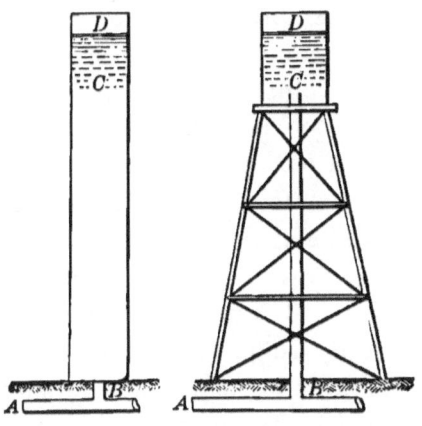

STAND PIPES.

stand pipe is to preserve pressure rather than to maintain storage, it is clear that height is an element of greater importance than diameter. The water in the lower portion of the stand pipe is, moreover, not available for storage, since the water level cannot fall to the base without causing all pressure to vanish. Hence the form of construction shown in the right-hand diagram of the figure is frequently adopted; here a trestle tower is built, on which the stand pipe CD is erected, and the water is carried to it by a vertical pipe BC rising from the main AB. The pressure produced at B by the head BD is the same in the two cases, but in the second one only the water in CD is available for storage.

The thickness of the plates of a tank or stand pipe generally increases from the top toward the base, since the higher pressure requires the greater amount of metal to resist it. The computation of thickness is exactly like that of a water main under a given head. If h be the head in feet the pressure in pounds per square inch is $p = 0.434h$. Then, as shown in treatises on applied mechanics, the equation $2St = pd$ applies to all questions of longitudinal strength. Here d is the diameter of the pipe and t the thickness of the metal, both in inches, and S is the tensile stress per square inch produced in the metal by the water pressure. For cast iron a safe value of S is about 2000 pounds per square inch; for a pipe of medium steel whose joints have an efficiency of 70 per cent a safe value of S is 9000 pounds per square inch. Thus in the figure if the head BD on the base of the steel stand pipe be 80 feet and its diameter 20 feet the thickness of the plates should be ½ inch, while at the mid-height the thickness need be only one-half as great. For the second diagram if the cast-iron pipe BC is 18 inches in diameter a thickness of ⅛ inch is theoretically sufficient to withstand the pressure, but cast-iron pipes must be made thicker than this on account of the stresses to which they are subject in transportation and handling. Tanks and stand pipes also receive stresses from the action of the wind and these must be carefully taken into account in the design.

Several failures of tanks and stand pipes due to the action of wind, to the accumulation of ice near the top, and to defective material or workmanship have occurred. The effect of a gale of wind on an empty stand pipe is often more injurious than on one that is filled with water; as additional security against wind tall stand pipes are frequently provided with guys of steel rope. A thick layer of ice at the top, falling after the water level has been drawn down, has been the cause of failures. One of the tallest stand pipes ever erected was that at Gravesend, N. Y.; this was 250 feet high, 16 feet in diameter at the base, and 8 feet in diameter at the top; at the first trial, when the water had reached the height of 227 feet, a crack occurred near the base and an instant later the entire structure fell with a deluge of water.

Stand pipes proper placed near the pumping engine are not now built as extensively as formerly, since the method of pumping directly into the pipes has been so developed that reliable regulation of the supply can be assured; the method of using a trestle base has increased in favor, and in some cases a wooden construction with iron hoops is employed. Large metallic tanks are extensively built on masonry and concrete foundations in order to obtain storage sufficient to allow stoppage of the pumps for a few hours. The height of water in the tank is known at the engine house either by readings of the water gage or preferably by an electric device operated by a float in the tank. A manhole is provided near the base of the tank, so that access may be had to it without climbing over the top when repairing or cleaning is to be done.

45. STREET MAINS AND FIRE SERVICE.

The first street mains used in the United States were made of logs, through each of which a hole 2 or 3 inches in diameter was bored, and these were connected with an end mortise-and-tenon joint. Cast-iron pipes were next used, and these are now more extensively employed than any other kind. Steel-riveted pipes cannot economically compete with cast-iron ones except for the large sizes required in conduit lines. Wooden pipes made of staves and bound with steel hoops have been used for street mains only in a few towns.

Cast-iron pipes are made of different thickness, depending on the diameter and the head under which they are to be used. For instance, pipes 6 inches in diameter have thicknesses of 0.41 and 0.45 inches for heads of 100 and 200 feet respectively, while pipes 12 inches in diameter have thicknesses of 0.53 and 0.60 inches, and pipes 24 inches in diameter have thicknesses of 0.76 and 0.90 inches for the same heads. They are cast in lengths of 12 feet, each having a spigot end and a bell end. When laid in the trench the spigot end of one length is inserted into the bell end of the next length, a gasket is forced into the annulus to the proper depth, and melted lead is poured in to fill the joint. Whenever a change in diameter is required this is made by a special tapering length called a reducer. Special lengths shorter than 12 feet are also provided for curves, for points where branch pipes are to connect, and for places where stop valves are to be inserted.

Hydrants are usually placed at street corners; in towns the post hydrant is most common, while the flush hydrant is used in cities, as the latter form does not project above the pavement. In the northern states a frost casing is necessary to prevent freezing. The connection of the hydrant to the main ought to be by a pipe not less than six inches in diameter in order that the pressure in the main may not be lost by the friction of the flowing water. The valve that closes the hydrant should move slowly in order to prevent the water ram that occurs under a sudden closure.

When a high pressure exists at the hydrants two or three lines of fire hose may be attached to one, and thus streams may be thrown without the help of fire engines. If the pressure at the hydrants be 80 pounds per square inch or more it may be called high pressure. When one line of best hose, 50 feet long and having a 1-inch smooth nozzle, is attached to a hydrant where the pressure is 80 pounds per square inch the discharge at the nozzle will be about 245 gallons per minute and the height to which the stream will rise for effective work will be about 85 feet, although some of the water will rise to 130 feet. If two or three lines of hose be attached the discharge of each and the effective heights of the streams will be much lower, since the increased draft will cause a decrease in pressure at the hydrant.

The following figures, derived from the experiments of Freeman, show the results that may be expected from different hydrant pressures by the use of 100 feet of ordinary best quality rubber-lined hose with a 1-inch smooth nozzle attached:

Pressure in pounds per square inch	20	40	60	80	100
Discharge in gallons per minute	117	167	205	236	263
Vertical height of stream in feet	27	52	72	82	89
Horizontal range of stream in feet	31	48	60	70	76

The heights and ranges here given are those at which the jet will be a good effective fire stream when a moderate wind is blowing, the extreme drops going considerably further. With common hose these distances will be somewhat decreased; and, of course, the longer the hose the less will be the discharge and height of the stream. The great value of pressure in fire service is very plainly shown by these figures, and a town which has a gravity supply with high pressure enjoys a good protection without the expense of maintaining fire engines.

High pressure, however, like many other good things, has some disadvantages. One is that it requires the street mains and house-pipes to be of greater thickness, and hence more costly, than under a medium pressure. Another is that it increases the consumption, for the greater the pressure the greater becomes the waste due to leakage and to carelessness of consumers. When part of a town is on a hill the maintenance of a high pressure there involves an excessive pressure in other parts unless some method of reducing it is employed, and for this purpose pressure regulators are often used.

A pressure regulator is an apparatus inserted in a pipe line which reduces the pressure in the pipes below it to such a limit as may be desired. The principle of its action will be understood from the figure, where A represents a pipe in which the pressure is 100 pounds per square inch. This pipe runs into a chamber where the pressure acts against a piston loaded with a weight W.

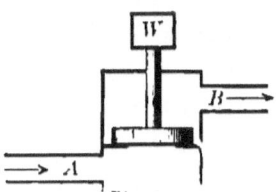

PRESSURE REGULATOR.

This weight is sufficient to cause a pressure of 40 pounds per square inch on the lower side of the piston, and hence the pressure in the pipe B cannot exceed 60 pounds per square inch. If water be drawn out anywhere along this pipe B the pressure in it falls below 60 pounds per square inch, and hence the piston rises and water flows from A into B until the pressure is restored. By varying the weight W the difference of the pressures in the two parts of the chambers may be regulated at pleasure. Instead of a weight a spring is generally used, or sometimes a weighted lever.

When the direct system of pumping is employed the regulation of the pressure within certain limits may be made at the engine house. During the night it may be lowered in order to prevent waste, and it may be quickly increased to the maximum limit by the man in charge when a fire breaks out in the town. In order that no failure in this program may occur it is necessary that an emergency should always be anticipated and that constant and vigilant foresight should be exercised to meet it.

46. Water Meters and House-pipes.

The measurement of the daily consumption of a town is effected in a gravity system by noting the water levels in the distributing reservoirs at times when there is no flow into them; these levels together with known areas of the water surfaces enable the volume taken out of the reservoirs to be computed. In a pumping system the consumption is determined from the displacement of the pump cylinders and the number of strokes. There is also an apparatus, known as the Venturi water meter, which may be placed in a pipe line and which will make a continuous record showing the consumption during every minute, hour, and day.

Water meters for the measurement of the consumption in factories and hotels are frequently used, and the system is in some cities also applied to many dwellings, as it is found that the waste of water is thereby much lessened. Commonly a consumer pays by the year, but in the meter system he pays by the gallon and accordingly faucets are not left open unless it is necessary. Wherever meters have been introduced it has been found that a marked decrease in consumption has resulted. For example, at Hoboken, N. J., in 1883 there were 2700 taps, of which 47 had meters, and the mean daily consumption was 121 gallons per person; in 1888 there were 5600 taps, of which 2667 had meters, and the mean daily consumption had fallen to 55 gallons per person.

A house water meter is, like a gas meter, provided with dials which register the consumption in cubic feet or gallons. The pointers on these dials are attached to wheels and these are turned by parts of the apparatus which move when water flows through. The piston meter is one of the most reliable kinds, and in this the motion of water causes two pistons to move in opposite directions, the water entering and leaving the cylinders by slide valves somewhat similar to those in a steam engine. The rotary meter has a wheel incased so that it is caused to move by the water passing through. The screw meter has a helical surface which revolves on its axis as the water enters at one end and passes out at the other. The disk meter has a wabbling disk so incased that its motion is proportional to the volume of water passing through. All meters require to be tested before being put into use, so that the error of registration, if any, may be known.

The house supply is brought from the main by a wrought-iron pipe usually 1 inch in interior diameter, but the connection to the main is generally made by a tap not greater than $\frac{3}{4}$ inch in diameter. This tap is inserted by the water company, but the consumer lays the pipe into the cellar of his house, where it passes through the meter, when one is used, and then branches to the laundry, the kitchen, and the bathroom. All these branch pipes should be of wrought iron, for lead pipes sometimes act unfavorably upon the water and holes may be eaten in them by rats. The pipes should be everywhere visible, except when it is necessary to run them horizontally under a floor, in order that ready access to them may be had whenever alterations or extensions are needed. They may be bronzed or galvanized in the kitchen and bathroom, or for these rooms brass pipes may be used by those who can afford the increased expense.

In the kitchen the water is carried into a vertical hollow cylinder called a boiler, and from the bottom of this a pipe runs through the water-back in the range and back again. The hot water from the range rises to the top of the boiler, whence the hot-water pipes run to the laundry, the kitchen sink, and the bath-room. Whenever hot water is drawn cold water enters the boiler, sinks to the bottom, and circulates back again through the range.

Each water closet is provided with a flush tank which is supplied from a cold-water pipe; when the tank is discharged a float falls and opens a valve in the pipe, and when the tank is again filled the valve is closed by the rising float.

For a house of ten or twelve rooms there are in the laundry about six faucets, three for cold and three for hot water. The kitchen sink, the bath-tub, and two wash-stands have four faucets for cold water and four for hot water. Two water closets have two cold-water connections. All these fixtures are provided with overflows, so that the water cannot rise above a certain height. For this water the householder pays from $10 to $15 per year and may use and waste as little or as much as he desires. On the meter plan he pays from 10 to 15 cents per thousand gallons. For a family of ten persons and the generous daily allowance of 50 gallons per person the meter plan will generally be more economical than that of yearly payment.

The water-works are now completed and the supply is brought into the houses. The collecting reservoir has impounded the run-off and delivered it to the distributing basins, or the pumps have raised the river water from a well at the bank. The quality of the water has been improved by aeration, sedimentation, and natural filtration, or if of suspicious quality it has been purified either by artificial filtration or by mechanical methods. An abundant supply under ample pressure fills the street mains, giving full security against fire, and allowing all street pavements to receive thorough cleaning. Fountains rise in the public park, and every lawn is kept green in the time of drought. The faucets are opened in the houses, and out of them comes pure and sparkling water whose use brings strength and health to the family. The first part of the construction work of the sanitary engineer is done. But the clear and pure water is rendered immediately impure by its use, and hence the second part of his work is to follow, whereby the foul water or sewage is to be removed from the town in such manner as to still further promote the health of the community.

47. Exercises and Problems.

31. Consult Baker's Manual of American Water Works for 1897, and compare five or six cities with respect to population, kind of system, daily consumption, and method of filtration. See Flynn's statistics in Engineering News, July 9, 1898.

32. What pressure in pounds per square inch is produced by a static head of 230 feet of water?

33 (a) A reservoir is full on July 1, receives 600 000 gallons per day during July, 400 000 during August, and 200 000 during September. What should be its capacity in order to furnish a town with a supply of 700 000 gallons per day during these three months and yet be one-third full at the end of September?

33 (b) Solve the same problem, taking the mean area of the water surface as 2.85 acres and supposing that the evaporation during the three months from this surface is 0.1 inches per day.

34 (a) An earthen dam has a width of 18 feet on top, a front slope of $1\frac{1}{2}$ to 1, and a back slope of 2 to 1. How many cubic yards of

material are required for a height of 36 feet and a length of 100 feet? how many for a height of 24 feet and a length of 100 feet?

34 (*b*) The South Fork dam near Johnstown, Pa., had an area of 407.4 acres at the ordinary height of water and 456.8 acres when the water level was 5 feet higher. The mean discharge of the waste-weir while the water was rising through this 5 feet was about 1000 cubic feet per second. If the rainfall was 0.8 inches per hour and one-half of this reached the reservoir, how many hours were required to cause the rise of 5 feet?

34 (*c*) Consult the novel "Put Yourself in His Place," by Charles Reade, and give his description of the breaking of the dam of Dale Dyke reservoir near Sheffield, England, in 1861. Ascertain the main dimensions of this dam and the cause of its failure.

35 (*a*) A masonry dam 36 feet high has a vertical back, its top thickness is 7½ feet, and its cross-section is a trapezoid. Compute the proper base thickness.

35 (*b*) A masonry dam of trapezoidal cross-section is 60 feet high, its top thickness is 9 feet, its base thickness is 30 feet, and the batter of the back is 2 inches per foot. Find where the resultant pressure V is applied on the base of the wall, and state whether the dam has the proper degree of security.

35 (*c*) Consult Schuyler's monograph on Reservoirs for Irrigation, in Report of U. S. Geological Survey for 1896–97, and describe the construction of rock-fill dams. Describe also the Bear Valley dam and the Sweetwater dam.

36 (*a*) A watershed of 2.64 square miles furnishes water to a reservoir whose area is 3.25 acres. If a rainfall of one inch per hour occurs and one-half of this reaches the reservoir in one hour, what should be the width of the waste-weir if its sill is 4.5 feet below the top of the dam?

36 (*b*) Consult Transactions of American Society of Civil Engineers for 1891, Vol. XXIV, pp. 431–469, and give further particulars regarding the failure of the South Fork dam.

37. Consult Report of the Aqueduct Commissioners of New York City for 1887–1895; give an account of some defects in construction of the new Croton Aqueduct due to inefficient inspecion, and explain how these defects were remedied.

38 (*a*) A pipe line represented by the figure is 23 700 feet long and its inner diameter is 12 inches. The elevations of the water

levels A and B above tide-water are 694.3 and 587.3 feet, and that of the point E is 597.5 feet; the distance along the pipe line from E to B is 10 350 feet. Find the pressure in pounds per square inch at E when the valve at A is open and that at B is closed. Find the pressure when both valves are open.

38 (*b*) Compute the diameter required to deliver 6 000 000 gallons per day through a pipe 18 320 feet long with a total fall of 13 feet.

39 (*a*) A distributing reservoir about 15 feet deep has an area of 20 574 square feet at the highest water level, 16 175 square feet at 5 feet and 10 440 square feet at 10 feet below that level. How many gallons are contained in the upper 10 feet of the reservoir?

39 (*b*) Consult Humber's Water Supply of Towns and Cities (London, 1876) and Fanning's Water-Supply Engineering (New York, 1890). Give sketches illustrating the construction of embankments and gate chambers for distributing reservoirs.

40. If the water gage at a pump reads 65 pounds per square inch, and 250 000 gallons of water are pumped in one hour with a suction lift of 6 feet, what is the effective horse-power of the pump?

41 (*a*) Visit a pumping engine and describe the boilers, steam cylinders, and water cylinders; ascertain steam and water pressures, height of suction lift, and capacity of the pumps.

41 (*b*) A pumping engine is to be purchased to deliver a mean daily consumption of 4 000 000 gallons, and the total lift, including frictional losses, is 72 feet. Taking the prices of engines of different duties as 10 per cent greater than those given in the table of Art. 41, and coal at \$4 per ton, determine which engine is the most economical?

42. In the figure the pipe AC is 180 feet long and 18 inches in diameter, while CB is 1800 feet long and 12 inches in diameter. If the total height of lift from A to B is 105 feet, what horse-power is required to pump 150 000 gallons per hour?

43. Collect data regarding the water supply of Indianapolis, Ind., St. Louis, Mo., New Orleans, La., and other cities.

44 (*a*) Consult the articles by Pence in Engineering News, 1894, and describe the failures of the stand pipes at East Providence, R. I., Peoria, Ill., and Thomasville, Ga.

44 (*b*) A steel stand pipe 180 feet high and 12 feet in diameter has plates $\frac{1}{2}$ inch thick at the base. What is its factor of safety?

44 (c) Visit a steel tank or stand pipe, take its dimensions, and make drawings of the horizontal and vertical riveted joints. Explain why the rivets are differently arranged in these joints.

45 (a) Collect sketches showing the arrangement and operation of the valves in different kinds of hydrants.

45 (b) What are the theoretic heights to which a stream will rise under pressures of 40 and 80 pounds per square inch?

46 (a) Collect sketches showing the arrangement and operation of the Crown meter, the Thomson meter, and the Worthington meter.

46 (b) Make a vertical section of a kitchen boiler and show how the water circulates in it.

47 (a) What is electrolysis in water mains? What is the distinction between a ring nozzle and a smooth nozzle? What is the derivation of the word Plumber? Explain the action of an automatic flush tank for a water closet.

47 (b) A water company is assured that, by extending its mains to a neighboring village, 400 taps may be obtained, each yielding $7 per annum. If money can be borrowed at 5 per cent per annum, and a sinking fund at 3.5 per cent compound interest be established to repay it in 12 years, what sum is the company justified in expending in order to earn during those 12 years a net income of $200 per year?

Chapter IV.

SEWERAGE SYSTEMS.

48. Historical Notes.

The organic wastes of a household are of two kinds, garbage and sewage. Garbage is the solid refuse of kitchens, such as vegetables, pieces of meat, and bones. Sewage is the liquid refuse of the laundry, kitchen, and bath-room, and consists of water fouled with soap, vegetable and animal matter, urine and fæces.

The ancient methods of disposing of these household wastes, and the methods which are still followed by the larger part of mankind, were to throw them out upon the ground to evaporate and act as manure, to bury them in the ground, to cast them into a stream, or to consume them by fire. Earth, air, water, and fire are the four natural deodorizers and purifiers of decaying organic matter. Air and water furnish oxygen, which enables the bacteria to perform their useful work of decomposing such matter into harmless constituents; earth acts in connection with air to produce the purification of liquid wastes in the same manner in which water is purified by filtration; fire directly oxidizes or consumes both the decaying garbage and the noxious bacteria which accompany it. Indeed all modern methods for the disposal of garbage and sewage depend upon the scientific application of these properties of earth, air, water, and fire.

Savage and barbarous man throws his refuse into a pile near his tent or hut, and when the collection becomes so large as to cause offense and disease he moves his dwelling to a new location. Civilized man removes the refuse at once from the vicinity of his dwelling and thus keeps the air and soil around him free from pollution.

A sewerage system is a plant for the removal of sewage from a town. The water-supply system brings to the town pure water, the sewerage system carries away the same water fouled with household wastes. The pure water enters the town through a single pipe line, which branches into the street mains, and these again branch into the house-pipes. The foul water or sewage leaves the houses through small drains connecting with larger ones in the streets, and these unite into one large sewer which conveys the sewage to a safe distance from the town. A sewerage system hence requires a water-supply system for its successful operation.

Damascus, the oldest city of the world, and regarded by the ancients as the fairest of earthly paradises, has always had an abundant supply of pure water from the river Abana, which also carries away the refuse from both streets and dwellings. At Nineveh and Jerusalem arched drains or sewers were built in early times. At Rome a great sewer called Cloaca maxima was built in 558 B.C. to drain the valley of the Forum; this sewer was 32 feet in height and about 500 linear feet of it are still in service. Later Rome also built many sewers to carry away the water furnished by its extensive system of aqueducts, so that Pliny speaks of it as a city on arches.

These ancient sewers, like many modern ones, served for two purposes: first, to carry away the rainfall and the run-off of brooks; and second, to carry away the foul water or sewage proper. An underground conduit which is built merely for the first purpose is generally called a drain; when sewage is admitted into it the word sewer should be used. A drain is usually of rude construction, often made by walling in and arching over a brook; a sewer, however, must be water-tight, so that the soil may not be polluted by leakage. London had drains during the seventeenth and eighteenth centuries, but no household refuse except liquid kitchen slops was allowed to enter them until 1815. The real sewerage system of London dates from 1847, when the drains had been so improved and extended that it was made compulsory to turn all sewage into them.

Prior to 1850 the methods for removing the sewage of towns and cities were generally the same as those still in use in villages and country districts, namely, kitchen sewage was run into the streets and fields to evaporate or percolate into the ground, while the sewage of privies and water closets was run into vaults and cesspools either to soak into the soil or to be removed at stated intervals. Both the privy vault and the cesspool were intended to allow the sewage to remain upon the premises as long as possible, and they were emptied only when the offense became great. As a consequence the soil of towns and cities was fouled, wells were polluted, and epidemics of disease were caused. By the use of water-tight vaults and cesspools, which are emptied only in cold weather, this system may be made a good one in country districts, but in large towns and cities it has proved to be impracticable from a sanitary point of view.

A pail system for the removal of the refuse of privies and water closets was in use in a number of European cities at the beginning of the nineteenth century and is still employed to a slight extent. This system consists in placing pails or tubs beneath the seats of privies and removing them at weekly intervals, both the vessel and its contents being taken away and replaced by a clean vessel. Although it might be thought that this method is an improvement over the privy-vault or cesspool system, the facts show that the offense caused by the weekly removal of the material is so great that no American town or city would tolerate it. The earth closet, introduced about 1865, was a modification of the pail system in which the vessel was partially filled with dry earth, and more added from day to day until it became full; then it was carried away and the contents used on the fields as manure. This method also proved impracticable on a large scale by reason of the trouble, offense, and expense which its operation involved.

The water-carriage method of the ancient Romans has proved to be the only practicable and economical one for the removal of the sewage of large towns and cities. This method, since its development in London and Paris, has rapidly spread over Europe and America. As previously remarked, a water supply is indispensable for its operation. In the United States there were in 1898 about 4000 towns and cities having a good water supply, but probably not one-fourth of these had an efficient sewerage system. The reason for this lies in the fact that a water-works brings a direct financial return, while a sewerage plant apparently does not. Private companies will build and operate a water supply for a town and secure a revenue that yields a good profit on the investment, but the construction of a sewerage system must be done by the municipality, and the average taxpayer sees only the expense and is not able to appreciate its benefits. These benefits, as set forth in the first chapter, are really nearly equal to those of the water supply, but often they are only fully appreciated when unclean streets and cesspool pollution produce an epidemic of disease. Such object-lessons have been so numerous that the more intelligent citizens of all large towns generally recognize the advantages of sewerage and advocate its introduction. The work now done by boards of health in investigating the causes of epidemics and in suggesting methods for their prevention, is exerting a most important influence upon the public at large. Undoubtedly long before the close of the twentieth century every town which has a water supply will be provided also with efficient sewerage.

49. House Fixtures.

About one per cent of the pure and abundant water supply which enters the house is used for drinking and for the preparation of food. The remainder falls into the laundry tubs, the kitchen sink, the wash-stands, the bath-tub, and the water closet, whence it immediately runs away as sewage. All these fixtures should be arranged to secure perfect cleanliness; they should not be boxed in with woodwork, but every part be left exposed to view, so that there may be no place around them where thoughtless servants can conceal sweepings or dirty rags.

Each of these fixtures must be provided with a trap, so that a seal of water may be preserved in its outflow pipe. This is necessary because the main soil pipe of the house sometimes becomes filled with foul gas from the decaying sewage, which would rise into the rooms through the outflow pipes were it not for the water seal of the trap. Of the many kinds of traps only four will be mentioned. The bell trap,

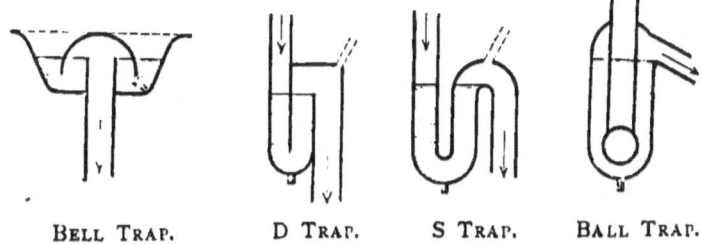

BELL TRAP.   D TRAP.   S TRAP.   BALL TRAP.

sometimes used in sinks, has a bell attached to the strainer to prevent the rise of gases; this is an objectionable trap because the space around the bell is so large as to collect solid matter, and moreover its efficiency is entirely destroyed whenever the strainer is removed. The D trap is a better arrangement for a sink, and the S trap is better still. The last sketch shows one of the so-called mechanical traps where a rubber ball is held by the pressure of the water against the end of the inlet pipe; other mechanical traps use valves or floats for the same purpose. There are also traps in which a mercury seal is used, arrangement being made that the water cannot force out the mercury.

A good trap is one constructed so that it will be thoroughly scoured by the water passing through it, so that evaporation cannot occur and so that syphonage cannot take place. The term syphonage is used to designate the emptying of the trap; thus if water moves slowly through the S trap the water level will remain as shown in the figure, but if a large quantity is discharged the velocity causes a negative pressure, so that the atmosphere acting down the inlet pipe forces all the water out of the trap. To prevent syphonage a vent pipe, shown by broken lines, is inserted on the sewer side of the trap and this runs to an open vent tube; by this arrangement the atmospheric pressure acts on the water in both sides of the trap. It is an advantage of some of the mechanical traps that syphonage cannot occur, and hence that vent pipes and tubes are unnecessary. It is desirable that every trap should be so arranged that it can be drained and its interior be inspected by the plumber.

The laundry tubs discharge soap and the kitchen sink discharges both soap and grease, which are liable to clog the main soil pipe. In hotels and large houses this evil becomes so great that a special device, called a grease trap, is generally used. This is a box placed below the main trap and having a cover which can be easily taken off. The inlet pipe enters near the top, and the outlet pipe is a syphon rising from the bottom, and thus the soap and grease are left to float on the surface of the water, whence they are removed once a week.

Every laundry tub, sink, wash-stand, and bath-tub must be provided with an overflow pipe, and it is important that this should join the outlet pipe above and not below the trap, for otherwise the gas will rise into the room through the overflow pipe. It often happens that plumbers, in order to effect a straight connection, will make such an improper junction. If possible all traps should be above the floor, where they are easy of access, and by a little careful planning on the part of the plumber this can generally be done. These overflow pipes, like the discharge pipes of the fixtures, are commonly of lead, with wiped joints.

The oldest style of water closet, called the pan closet, had a pan operated by a lever, and the contents of the pan were dumped into a bowl beneath, which soon became foul and offensive; happily this dangerous form has now gone out of use. Later the plunger closet, the wash-out closet, and the hopper closet were introduced, and of each of these there are

PLUNGER CLOSET.   WASH-OUT CLOSET.   HOPPER CLOSET.

many kinds and styles. The plunger closet is a modification of the old pan closet in which the bowl is omitted, and a large body of water is released when the plunger is raised. In the wash-out and hopper closets the discharge of the water tank by pulling the chain gives an ample amount of water to effect thorough flushing. The hopper closet appears to take the highest rank on account of the simplicity of its construction, and numerous styles of it, called the short hopper, the long hopper, the wash-down, and the syphon-jet closet are found in the market. The water is admitted in all cases around the rim, but some styles also bring in a jet lower down. To avoid syphonage of the trap an opening is provided to which a vent pipe is attached, this pipe running to the vent tube.

A closet is made of earthenware with a porcelain glaze and

so built that no portion is below the floor line. No woodwork of any kind, except the seat and lid, is needed around it when placed in a bath-room. It should be placed near the main soil pipe, so that the connection between them is short, and it is important that the waste pipes from the bath-tub or washstands should never connect to the discharge pipe of the closet, but be carried independently to the soil pipe.

In hotels and public buildings the closets are necessarily placed in stalls, and these are often made of wood, although marble or slate is used in the best work. No stalls for urinals, however, should ever be made of wood. In spite of all these precautions a certain degree of attention on the part of a servant is needed to secure perfect cleanliness.

50. HOUSE DRAINAGE.

House drainage, or house sewerage, as it should properly be called, embraces all those pipes and connections which carry the sewage out to the cesspool or the sewer, the former being the destination for an isolated country residence and the latter for a city house. It consists of two parts, an iron pipe within the house which extends about three feet beyond the wall, and an earthenware pipe which extends through the ground the remainder of the distance.

The soil pipe is of cast iron, 4 inches in diameter, with lead joints. It projects above the roof of the house, with the top open, this being protected by a basket to prevent ingress of birds and rats. Through the house it is vertical until the cellar is reached, and then it curves so as to pass out of the wall horizontally. In the figure are shown two water closets connecting with the soil pipe at A and C and a kitchen sink at B; at D is a trap whose lid may be removed if an obstruction should occur. The soil pipe should be everywhere visible except at the places where it passes through the floors. It should not be carried under the cellar floor except when connections to laundry tubs or water closets demand it; in this case the trap *D* should be made easy of access by placing it in a brick box with a cover at the level of the cellar floor. These precautions render it impossible for any leakage to

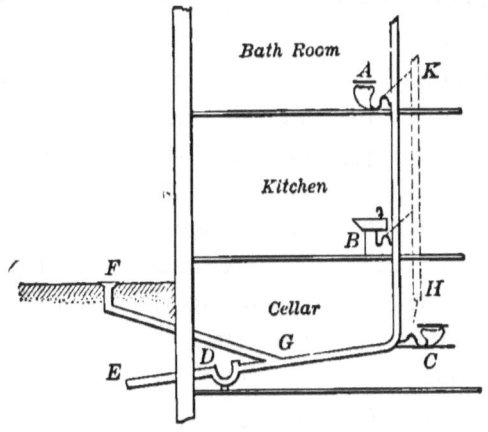

HOUSE-SEWERAGE PIPES.

occur without being soon detected, and make the plumbers' bill for repairs much smaller than if a wall or cellar floor has to be torn to pieces and rebuilt.

A fresh-air inlet pipe *FG* connects the soil pipe to the external air, so that a circulation may occur; this will certainly take place if the soil pipe be placed near the house chimney, whose heat will render the air in it higher in temperature than that outside the building. By this method oxygen is supplied to attack the decaying organic matter in the soil pipe and purify it by combustion and nitrification. Such inlet pipes may be placed on the outside edge of the sidewalk in cities, and no offense will be caused, since in all ordinary cases the air of the street is entering them in order to pass out again at the tops of the houses.

The vent tube, shown by the dotted lines at *HK*, is also iron and usually about 3 inches in diameter. This does not act in any respect as a ventilating flue, but its office is merely to bring atmospheric pressure on the crowns of the traps and thus prevent their syphonage. This vent tube may be joined to the soil pipe above the highest fixture, or it may be carried up above the roof; in the latter case it is well to make it 4 inches in diameter, so that clogging by frost or snow may not occur.

When the rainfall on the roof is admitted to the sewer the water leader runs down on the outside of the wall and discharges into the soil pipe near E. If the junction is made by a tight connection a second trap may be inserted beyond it, but it is often preferred that the rain water should discharge into a small manhole which is connected with the soil pipe; either of these arrangements brings the atmospheric pressure on the street side of the main trap D and thus renders it less liable to syphonage.

The second part of the house-drainage system is a vitrified-clay pipe, called the drain pipe, which extends from the end of the iron pipe to the sewer; this is 4 inches in diameter, or 6 inches for hotels and very large houses; it is laid with cement joints, and its slope should not be less than 3 feet in 100 feet. When this pipe goes to a cesspool, as in a country house, the cesspool should have a vent to the open air.

After the completion of a house-sewerage system, but before the vent and drain pipes are connected, a test for leakage may be made. This is done by stopping the end of the soil pipe and all fixture outlets and vent pipes, and then filling the soil pipe with water up to the top and allowing it to remain for twenty-four hours. This test cannot be made in freezing weather, and it is often an awkward operation to undertake. A better test is that by peppermint, which shows the efficiency of both traps and joints. This is made after the work is entirely completed and the traps filled, and consists simply in putting about half an ounce of oil of peppermint into the fresh-air inlet and noting whether any odor is observed in the house. In cases where there is no fresh-air inlet the peppermint may be put into the lowest house fixture and be quickly washed down with water, or it may be put into the cesspool or sewer. The man who handles the peppermint should not walk through the house until after the close of the test.

It was said forty years ago that a man's happiness was inversely proportional to the number of gas, water, and sewer pipes in his house. This saying may be true to-day if these modern conveniences be put into the house with the main idea of cheapness, without a definite plan, and without inspection of the plumbers' work. In many cities the law requires that plans and specifications for house drainage must be approved by the board of health or by the inspectors of buildings before work can be begun, that the plumber employed must be one licensed by the same authority, and that official inspections must be made. Under this plan the house-sewerage system will be one that brings comfort and happiness to the householder and promotes the health of the family and the community. The same result can be secured when plans and specifications are drawn by a competent architect if these are followed by a strict inspection of the work of the plumber. Here, as in all other branches of sanitary engineering, well-laid plans and constant inspection in executing them are indispensable in order to secure health and happiness.

51. CLASSIFICATION OF SYSTEMS.

In Art. 48 it was shown that the pail method for the removal of excremental matter is an offensive and impracticable one, and, as it is not used in the United States, it will not be classed among the public systems. The cesspool plan was also shown to be an objectionable one for a large town or city, and, although it is still extensively used in villages and country districts, it is to be regarded as a family method rather than a practicable and efficient public system and hence will receive no further consideration. There remain, then, only the systems of removal by means of the public water supply, known as the water-carriage systems, which are an outgrowth of the plan followed in ancient Rome. Had these ancient methods been continued and developed throughout Europe the thousand years of filth, disease, and misery known as the dark ages might perhaps have been a thousand years of cleanliness, health, and happiness.

In the two preceding articles the arrangement of the sewerage fixtures and drains of a modern house has been described, and now the sewage is to be carried by the street sewers through and away from the town. This sewage is mostly water and the amount of it is practically the same as that of the water supply that enters the house. The total amount of sewage carried away from the houses may in summer seasons be slightly less than that of the total water supply, owing to the loss by evaporation and percolation of that part used for sprinkling lawns and streets, but it is a fair assumption to take the two amounts as equal.

The total solids in an analysis of sewage may be stated as about 1000 parts per million for an average figure, and rarely if ever do they exceed 5000 parts per million. Hence much more than 99 per cent of the sewage is water, and all the laws of hydraulics are directly applicable to its flow through the sewer pipes. The oxygen in the water tends to decompose and destroy the organic matter, but the supply of it is only sufficient for a very imperfect purification, and hence it is important to remove the sewage as quickly as possible out of the town to a place where an abundant amount of oxygen is available. This place, in all the ancient sewerage plants and in the majority of the modern ones, is the ocean or the river, where the dilution of the sewage with water furnishes sufficient oxygen to enable the useful bacteria to complete the work of turning the organic matter into harmless substances.

51. CLASSIFICATION OF SYSTEMS.

The water-carriage systems for sewage removal may be divided into two classes, gravity systems and pumping systems. A gravity system is one where the flow of sewage takes place entirely by the force of gravitation in the same manner as that of water in a conduit or pipe; probably more than 90 per cent of all sewerage plants are gravity systems and they are always used when there is sufficient fall from the town to the place where the sewage is to be delivered. A pumping system is one where some method of lifting the sewage is required either on account of the low location of the town or in order to raise it so that it may be purified by the methods which are to be described in the next chapter. In a pumping system there is always a certain amount of gravity action, because it is necessary that the flow from the houses must be collected in wells before it can be pumped.

Gravity systems are divided into two kinds, called the combined system and the separate system. In the combined system the sewers carry not only the sewage, but also the rain water which falls on the roofs and the streets; this system is hence a close imitation of the method of ancient Rome. In the separate system the sewers carry the house sewage, while the water of the streets is entirely excluded; a small amount of roof water may, however, be admitted in order to assist in flushing the pipes.

A pumping system carries away sewage, which is collected in wells either by the combined or separate method, by raising it to the required height to insure the flow out of the town. Common pumps may be used for this purpose, but distinctive pneumatic methods have also been introduced. The vacuum system is a suction method by which a partial vacuum is formed in many pipes of a district so that the atmospheric pressure may lift the sewage. The compressed-air system is a method by which air pressure is transmitted through a series of pipes to vessels where the sewage has accumulated and thus forces it up to a higher elevation.

The combined and separate systems, without pumping, are those which should receive the main attention of the student on account of their more extensive use. It is apparent that the combined system requires larger sewers than the separate system, and also that when the latter is used extra provision must be made for disposing of the storm water of the streets. In some cases this storm water may be left to take care of itself by natural flow in the streets, but in others a special system of drains may be necessary.

The main use of a sewerage system is, of course, to remove the sewage out of the town before it has had time to decompose, but it also furnishes an important incidental advantage in draining the soil. The evil effects of wet soil and damp cellars are well known, and the sewers should be so constructed as to lower the level of the ground water in damp localities. This drainage cannot enter the sewers, for they should be made water-tight in order to prevent the pollution of the soil, but the sewer foundation can be so arranged as to act as a drain, and thus to greatly decrease the dampness of cellars and basements.

52. THE COMBINED SYSTEM.

The sewers of the combined system are to carry away not only the sewage, but also the rain that falls upon the roofs and streets. Hence the rain-water leaders are directly connected to the house drain pipe, and at the street corners basins are constructed to receive the flow of the gutters. The size of the sewers in this system depends more upon the storm water than upon the volume of the house sewage. The amount of sewage may be regarded as the same as that of the water supply, say 100 gallons per person per day as the mean amount, and 180 or 200 gallons per person per day as the maximum flow on Monday forenoons. The amount of storm water is more difficult to estimate, but a common method is to take one inch of rainfall per hour over the area covered by the sewerage system and to consider that the main sewer is to carry either all or a part of this water. As noted in Art. 16, rainfalls much heavier than this are liable to occur, but these come at rare intervals and are of short duration, so that probably one inch per hour is a fair maximum allowance for sewer capacity. To provide for an excessive rainfall of 4 or 5 inches per hour will involve an unwarranted extra expense, the interest on which would be more than enough to cover the cost of street repairs rendered necessary by heavy storms of rare occurrence.

In the design of a combined system the first thing is to prepare a topographic map of the town and its vicinity. This map shows the watershed whose storm water is liable to reach the sewer, all streets and houses, and the contour curves. For a flat town contours at vertical intervals of one foot are needed and the levels by which these are determined must be run with great precision; for a hilly or sloping town contours at intervals of two or three feet may perhaps be sufficient. With this map in hand the engineer is able to make profiles of streets, ascertain the available slopes, lay out the lines for the main and lateral sewers, and then estimate the cost of construction. It may be often necessary to make several plans in order that comparative estimates may determine the one which will furnish the proper efficiency with the greatest degree of economy.

The directions of the main and lateral sewers will be determined by the topography of the town. The simplest case is that where the town is on one side of a river with a uniform slope toward it; here there may be one or two main sewers running directly into the river, and this is called a perpendicular method of location. The more usual case is one where the main sewer runs parallel with the river and the lateral sewers run into it; this is called the intercepting method, because the main sewer takes the place of the river and carries

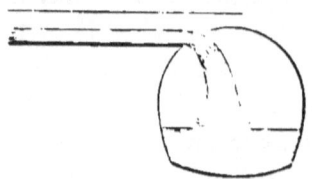

OVERFLOW CONNECTION.

all the sewage to some point down the stream. In order to relieve the main sewer of a portion of the storm water the arrangement in the figure is sometimes used; here the normal flow of sewage is caught, but when the lateral sewer becomes filled after a rain a portion of its flow is discharged over the top of the main sewer into the river.

The shapes of sewers of the combined system are circles for the small sizes, circular or egg-shaped sections for the medium sizes, and basket-handle sections for the large sizes. Circular sections are made of brick when over three feet in diameter,

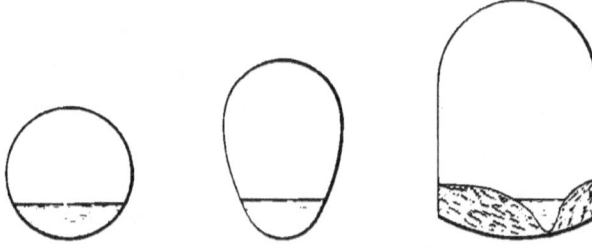

SECTIONS OF SEWERS.

while vitrified clay pipes are used for smaller sizes. Egg-shaped sections have the advantage of giving a small wetted perimeter under partial flow and hence of securing a velocity which prevents the deposit of organic matter. The two kinds of basket-handle sections are used for stiff and soft soil, as explained in Art. 37; these forms offer much frictional resistance to the flow under normal conditions, as the section of sewage alone is small compared to the area of the entire cross-section. The figure shows what sometimes occurs in one of these large sewers when the slope is slight, a large deposit of solid matter being made and the small quantity of sewage flowing in a channel formed upon it. During a period of prolonged drought these deposits may accumulate, unless hydrant water be turned in to remove them, so as to pollute the air of the surrounding neighborhood. One strong objection to the combined system is, in fact, the difficulty of keeping the large sewers in a cleanly condition during dry seasons.

Manholes must be provided at all points of junction, and also at regular intervals along the lines, in order that access to the sewers may be had. A manhole also serves as a ventilator, it being covered with a perforated iron plate, through which air comes out and enters. Indeed the only practicable method of ventilating sewers is by means of such manholes; and it is not found that the air of a good sewer causes any offense in the streets. The constant admission of fresh air furnishes oxygen to the decomposing matter and thus secures deodorization and purification. If the sewers were unventilated and fresh air not admitted an interior pressure might result, which would force sewer gas back into the soil pipes of the houses and so produce evil effects. Sewer gas in a house is most injurious, but in the street, where a constant supply of fresh air is at hand, it is quickly neutralized and becomes harmless.

The catch basins which receive the storm water are located at the street corners so as to receive the flow of two or more gutters. The pipe connecting with the sewer enters at one side sufficiently high so that sand and gravel may collect in the bottom, whence it is removed after the storm has ceased. Rubbish may be kept out of the pipe to a certain extent by curving it down like a syphon or by placing a basket screen over its end.

In seacoast towns, where the main sewer discharges below high tide, it is provided with a flap valve at the end to prevent the ingress of the water, and thus the sewage is backed up for some distance until the tide falls. Sometimes large tanks are built to receive this accumulation and store it for the three or four hours during which the outlet is covered by the tide.

The combined system is best adapted to large cities on flat ground where the storm water of the streets may cause great damage by flooding basements and sidewalk vaults. It is also well adapted to towns where the storm water cannot be otherwise diverted so as to prevent similar damages. In many towns the combined system has been a slow growth, first starting by the construction of a sewer to carry the flow of a troublesome brook. As has been remarked before, there is nothing in any system which renders it economical or advantageous in all cases, but in each case the engineer is to make such plans as will best fit the local circumstances.

53. The Separate System.

The separate system originated in England, it being first proposed by Phillips in 1849 and recommended by him as a solution of the difficulties in London, where large drains had been built to carry off surface water only and hence were not well adapted to receive sewage. Accordingly he maintained that the proper solution of the question demanded that sewage should be carried in a new system, "distinct and separate from the permeable land drains." The idea was not carried out in London, but it received the approval of some engineers and about 1870 such sewerage systems were constructed at Oxford and a few other English cities. In America the first application of the system was by Waring at Memphis, Tenn., in 1880. As in the case of all new improvements, much opposition and criticism was made, but the cheapness and efficiency of the system soon overcame these objections, so that since 1885 hundreds of towns have been sewered on this plan.

The sewers of the separate system are to carry sewage only, but a small amount of additional water is admitted from flush tanks or from roofs to insure cleanliness. As a consequence the sewers are smaller than those of the combined system, being rarely larger than four feet in diameter, and the cost of construction is materially lowered. Circular sections are used almost exclusively and for all the common sizes these consist of clay pipes with cement joints.

All the remarks made in the last article regarding the influence of topography on the location of the sewers apply equally well to the separate system. The rainfall over the watershed area is not to be considered, however, as this is not to be admitted to the sewers. When roof water is taken into the sewers for the purpose of flushing this is done only at a few houses near the dead ends of the lateral lines, and only the houses designated by the engineer are allowed to connect their rain-water leaders with the drain pipes.

Manholes are provided, as in the combined system, at all junctions of lateral with main sewers and at other points on the line from 300 to 500 feet apart. Through these manholes ventilation takes place, and they also give access to the sewers so that obstructions may be removed. A special form of air

AIR INLET.   LAMP-HOLE.   HAND-HOLE.

inlet is likewise often used; this is cheaper than a manhole and serves the same purpose as far as ventilation is concerned; this consists of a vertical pipe having its open top at the street surface in a masonry box which is covered with a perforated iron lid.

The early sewerage plants on the separate system suffered much from obstructions caused by deposits or by matter accumulating at a defective joint. To meet this difficulty manholes were constructed at closer intervals, so that rakes could be introduced or jets from a hose be thrown into the pipes. Lamp-holes like that shown in the above figure are also built between the manholes; as these are used only at rare intervals, the cover caps are placed below the street surface and a record of their location made. Hand-holes in pipes may be likewise inserted between the lamp-holes and their location be recorded, but to use these the street must be torn up.

The sewers of this system are generally designed in such sizes that they will be about half full when the maximum flow occurs on Monday forenoon. Hence the capacity is amply sufficient to carry double the mean daily water consumption and also to allow for the future growth of the town. The ordinary flow will then make the sewers less than half full, and the additional water needed for flushing may be admitted either during the maximum or minimum flow. It has been found that reliance upon roof water alone is not satisfactory, and that regular periodic flushing must be generally made from special tanks. In fact some engineers allow no roof water at all to enter the sewers, and probably this is the wisest plan.

A flush tank is placed near the dead end of a lateral sewer and is so arranged that the entire contents may be quickly discharged into it. The admission of the water to the tank is so regulated that it may take several hours, a day, or even longer to fill it, and the discharge then occurs automatically. Several of these flush tanks are placed throughout the town and by proper regulation of the entering water they may be made to act separately or simultaneously, as experience may determine best. They are built of masonry in the street and resemble a manhole in general appearance. The water is brought by a small pipe from the street main and its faucet is so located that it can be reached by taking off the iron cover of the tank.

There are numerous styles of flush tanks, which may be classed as tilting, syphon, and mechanical tanks. A tilting

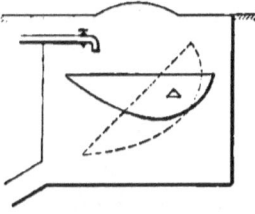

TILTING FLUSH TANK.

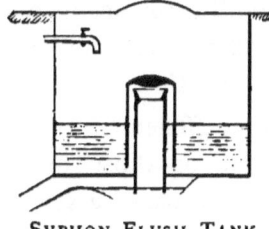

SYPHON FLUSH TANK.

tank receives the water in an iron box which is supported upon a knife-edge pivot at each end, the shape of the box being such that it remains horizontal until filled; as the water is admitted the center of gravity of the box and water approaches nearer and nearer to the pivot and finally when it reaches the pivot the box suddenly tilts and discharges its contents into the sewer. The syphon tank shown in the figure has a vertical pipe surrounded by a bell, the annular space between these constituting the syphon; as the water is admitted it gradually rises in the tank until the top of the syphon is filled and then runs over into the vertical pipe until the lower pool is filled; shortly after the water of the tank is suddenly discharged into the sewer. Mechanical traps depend upon floats or valves to initiate the discharge, and they are perhaps more liable to get out of order than those which depend only on gravity or atmospheric pressure. The tilting tanks usually discharge from 100 to 200 gallons, but the syphon tanks are of larger size and may discharge from 400 to 500 gallons in less than one minute.

54. SIZES OF SEWERS.

The flow of sewage in a sewer is in all respects controlled by the same laws which govern the flow of water in aqueducts and conduits. The sewer must be on a uniform slope, for the material of which it is made will not resist the interior pressure due to a head, and moreover such head could not exist without flooding the cellars along the line. The maximum flow in all cases occurs when the sewer is nearly but not quite full, and computations made by regarding it as full are hence on the safe side.

The coefficients of discharge for clean sewers may be taken the same as those given for aqueducts in Art. 37. The interior surface of a sewer, however, may become foul in time from deposits and incrustations, so that it is best to use the smaller values given in the following table. Here, as before,

COEFFICIENTS FOR SEWERS.

Hydraulic Radius in feet.	$s = 0.0001$	$s = 0.0020$	$s = 0.0004$	$s = 0.001$	$s = 0.01$
$r = 0.2$	68	74	78	81	81
$r = 0.4$	86	91	94	96	98
$r = 0.6$	95	100	102	104	106
$r = 0.8$	103	106	110	111	112
$r = 1$	109	113	115	116	117
$r = 1.5$	120	122	123	124	125
$r = 2$	127	128	129	130	130
$r = 2.5$	132	133	134	134	134

the slope s is the ratio of the fall in any distance to that distance, and the hydraulic radius r is the ratio of the area of the cross-section to its wetted perimeter. Thus if a sewer has a fall of 6 inches in 100 feet the slope is $s = 0.5/100 = 0.005$; if the area of its cross-section is 19.5 square feet and the inner perimeter is 16.8 feet the hydraulic radius is $r = 19.5/16.8 = 1.16$ feet; then from the table the coefficient is $c = 119$.

By the use of the above table and the formula $v = c\sqrt{rs}$ the mean velocity v is computed in feet per second, and then the discharge q follows from the formula $q = av$, where a is the area of the cross-section. In a case of design q is given, and here the values of a and r are to be found by trial from the equation $q = ca\sqrt{rs}$. For circular sections a solution can also be made by inserting for a and r their values in terms of the diameter d, and then $d = (8q/\pi cs^{\frac{1}{2}})^{\frac{2}{5}}$.

For example, take a town of 8000 people for which a combined system is to be designed, the area whose storm water is to be carried by the main sewer being 2 square miles. The maximum flow of sewage will be at the rate of 1 600 000 gallons per day, or nearly 2.5 cubic feet per second; the storm water, at the rate of half an inch of rainfall per hour, gives 645 cubic feet per second, an amount so large that the sewage itself is unimportant in comparison. By the use of the above method it is found that a circular sewer about 7 feet in diameter on a slope of 1 foot in 100 feet is needed, or the diameter should be about 11 feet if the slope be 1 foot in 1000 feet. The size is too large for a circular section, and hence a study and design of a basket-handle section must be made. Indeed, if this main sewer be very long and the slope slight its cost of construction might become too expensive for the town, and accordingly the engineer would be obliged to reduce the amount of storm water which it is to carry. The lateral sewers are then discussed in a similar manner and each made of such size as to dispose of the flow of its district.

In the separate system the sewers are to run only half full under maximum flow; hence $q = \frac{1}{2}av$, and the formula for circular sewers becomes $d = (16q/\pi cs^{\frac{1}{2}})^{\frac{2}{5}}$. For the above data, where $q = 2.5$ cubic feet per second, this gives a diameter of about 14 inches for a slope of 0.01 and about 20 inches for a slope of 0.001.

The slopes and shapes of sewers ought to be so arranged that the mean velocity of flow shall not be less than 2 feet per second, for if the velocity be smaller sedimentation occurs and deposits are produced which in time may become obstructions. Under a variable flow the egg-shaped section is a good one to prevent low velocity, because the hydraulic radius is larger for a small depth of sewage than a circular one. The basket-handle sections, on the other hand, are the poorest forms for a low depth of flow, as then the velocity becomes small and deposits may be formed in the manner indicated by the last figure in Art. 52.

The above numerical illustrations show clearly the great advantage of the separate system in cases where the storm water can be otherwise readily carried away. Small size means low cost, and hence it is that the separate system has been so widely adopted since 1885. Indeed, hundreds of towns which would now be using the old cesspool plan if the combined system of sewerage were the only one available have adopted and built an efficient system for the removal of house sewage on the separate plan. The enormous Cloaca maxima, 32 feet in height, through which Nero sailed in a stately boat, and some great sewers in London and Paris which may be also navigated by smaller boats, no longer serve as models, but they teach the lesson that more economical methods might undoubtedly have been adopted.

55. Construction of Sewers.

The design of a sewerage system involves not only the determination of the sizes, but detailed working drawings of all sections, junctions, manholes, basins, and outlets. When these plans have been prepared and the specifications drawn the work is let by contract to the lowest responsible bidder and the work of construction begins. It is not the intention here to discuss the hundreds of details which are involved in construction work, but merely to note a few points which involve general principles and have not been already mentioned.

In all soft and yielding soil secure foundations must be provided. A pile and grillage foundation is commonly used for both small and large sewers in wet soil, and on this masonry or concrete may be laid to form the base on which the sewer rests. In slightly yielding soil concrete alone may be sufficient, and in stiff soil the sewer may be laid in a bed of earth carefully excavated to the shape of the invert. As the exact character of the soil may not be known until the excavations are made, general plans for all kinds of foundation are pre-

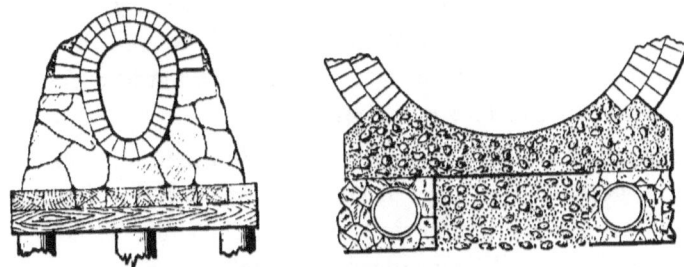

SEWER FOUNDATIONS.

pared in advance, and then each is built at the place designated by the engineer.

The drainage of the subsoil is to be provided for, when necessary, by stone or tile drains under the sewer, as indicated in the right-hand diagram of the above figure. The sewers themselves must be impermeable to water, for if water can enter sewage will leak out and thus cause pollution of the soil of the street.

In the combined system brick sewers are generally used in America for all sizes greater than about 3 feet in diameter. The thickness is usually 8 inches for the small sewers and 12 inches for the largest ones, the sides of the bricks being normal to the intrados curve and the additional width of the joint at the extrados being made by the mortar. The bricks break joints both transversely and longitudinally, and they should be wet before the hydraulic mortar is applied. A brick sewer is necessarily built on a wooden center formed of slats laid on cross-frames, this being pulled out and relaid as the work advances.

Concrete sewers are extensively built in Europe, and it is hard to say why they have not come into use in America further than the experimental stage. As they are built in place and thus form monolithic constructions, they should be superior to brick in strength and durability provided that the sewage does not have any deleterious action upon the concrete; it does not seem that this should be the case to any greater extent than upon hydraulic mortar. In the above figure a brick sewer with concrete base is shown, and this should be a good form if care be taken to avoid a continuous joint between the two materials.

The small lateral sewers of the combined system, and all but the very largest sizes of the separate system, are made by joining together vitrified clay pipes. These are usually 3 feet long and have bell and spigot ends which are connected by cement joints; in ordinary soil they need no foundation other than the bed of the trench. When passing under a railroad embankment, or in any place where shocks are liable to break the pipe, cast-iron water pipes should be used instead of clay ones.

The depth of a sewer below the street surface must be such that its crown is at least one or two feet below the level of the cellar floors in order that these may be effectively drained, and this generally demands a minimum depth of 8 or 9 feet. As the slope of the sewer is to be uniform throughout its length, it may be difficult to secure this depth in a street with irregular profile. A sewer is ordinarily built in a trench, but occasionally it is laid near the surface and covered with an embankment, and sometimes tunnel work is necessary.

All manholes, catch basins, and flush tanks are built on good foundations with stone or brick masonry. For the large sewers a manhole is an upward extension of the sewer itself, but for the small ones it is built independently and the clay pipes carried into and out of it at slightly different levels.

Branch hubs for the house drains are put on the sides of the sewers and above the middle at intervals of 25 feet along the line or opposite every lot where a connection may be hereafter made, and these are covered by a cap before the trench is filled.

The lines and grades for the work are given by the engineer and his assistants, who also exercise a constant inspection on the materials and workmanship and see that the plans and specifications are fully carried out. Books on construction give many details regarding brick, stone, cement, and their uses, but the greater part of the knowledge of the inspector is not and cannot be learned from books; it must be gathered in the stern school of experience. The construction and inspection of a sewerage system are not unimportant because it is buried in the ground as soon as it is built; but rather the most painstaking care and vigilant inspection should be exercised in order that no hidden defects may mar its successful operation.

56. Ventilation and Cleaning.

In the early days of sewerage systems, that is to say before 1870, much trouble occurred from the sewer gas forcing its way into the houses and creating offense in the streets. This was due to a number of causes: first, the house plumbing was imperfect and inefficient; second, the sewers were so large as to become very foul in dry seasons; and third, the manholes and air inlets were so few in number that sufficient air was not supplied to effect deodorization. When sewer gas did escape in a narrow street of a European city its volume was large and the offense was noticed at the windows of the houses.

In order to overcome these troubles many methods for the ventilation of sewers were tried in Europe. Tall chimneys were built near the ends of a sewer, and pipes were carried up on house walls along the line, to create a draft. Suction fans driven by windmills or steam engines were used to draw the air out of the sewers. The chimney plan did not usually produce the required draft, and the suction plan produced at some places a draft so great as to syphon all the house-traps and at other places an insufficient circulation. In time all these methods were practically abandoned, although a few chimneys may yet be found in Europe, and the fan method was very recently still in operation for a few of the sewers of London.

Various chemical expedients were also attempted to deodorize and neutralize the sewer gases before they escaped into the streets. Trays filled with charcoal powder were placed under the manhole covers. Chlorine and sulphur gases, iron carbides, and other substances were advocated and tried. All these methods likewise proved expensive and inefficient, and their use has hence been very limited.

The idea which mainly governed these early methods was the mistaken one that the sewer gas should not be allowed to escape into the streets. As soon as the falsity of this idea was recognized, and perforated manhole covers and fresh-air inlets were provided, the problem of sewer ventilation was solved. As soon as the house-drainage system was provided with proper traps, vent tubes, and inlet pipes the house was protected against the sewer gas. Some of the gas escapes into the streets, it is true, but the gas from fresh sewage is not more dangerous than that of the odor and dust from the animal excrements which are constantly dropped upon the pavements. If there be many of these openings the gas escaping from each is small in volume and the oxygen of the air deodorizes it far more efficiently than a large volume can be neutralized by artificial chemical treatment.

The gas from sewage in an advanced stage of decomposition is offensive and dangerous, and the method for keeping this out of the streets is to prevent its formation. If the sewage flows so rapidly that deposits cannot occur it will be removed from the town before the dangerous state of decomposition is reached, and to secure the necessary velocity of flow is one of the important problems in the design of the shape and size of the sewers. In the large sewers of the combined system this is often difficult, and even in the small sewers of the separate system some deposits are apt to form which may be the cause of obstructing the flow. Hence cleaning or flushing the sewers may be necessary from time to time.

In the combined system the main reliance upon flushing is the storm water, and when this comes the velocity is so increased that thorough cleaning is done. During a dry season, however, artificial flushing is often resorted to. This is done by building a dam in a manhole and arranging in it a sluice gate so that it may be raised from the surface. The sewage then backs up and hydrant streams are turned in at manholes above the dam; then when the gate is raised the rush of water scours out the sewer. Sometimes permanent dams with sluice gates are built in large special manholes, called penstocks, so that they can be put into operation with less expense than a temporary construction. If the sewer be a small one hydrant streams alone may often do much good work, particularly if the deposits can be stirred by rakes while the increased volume is flowing.

In the separate system the flush tanks described in Art. 53 are in daily operation and hence the sewers are always cleaner than those of the combined system. Such flush tanks can also be put at the dead ends of the lateral sewers of the combined system, but this is an uncommon practice. When an obstruction occurs in the sewers of the separate system manholes and lamp-holes enable the location to be closely determined, and if rakes and water jets fail to remove it the hand-holes are used as a last resort.

It is seen from these brief discussions that the operation of a sewerage plant is a matter which requires the careful intelligent supervision of the city engineer. The designing engineer does his best to render the sewers self-cleaning by making the slopes and sizes such as not to allow a low velocity of flow; the constructing engineer does his best to carry out these plans and to secure such workmanship as will prevent all obstructions; but yet maintenance cannot be neglected. The combined system has catch basins which must be cleaned and large sewers whose condition with regard to deposits should always be known. The separate system has flush tanks whose discharge must be regulated from time to time and manholes which must constantly be kept under inspection. In times of unusual rainfall the combined system may fail to carry the storm water and cause cellars to be flooded, and on occasions of serious obstructions the sewers of the separate system may back the sewage into basements through the drain pipes. To guard against all these contingencies and reduce their effects to a minimum the town or city engineer must exercise constant and vigilant care.

57. Pumping of Sewage.

When a town is very flat it may be impossible to carry away the sewage by gravity and hence some method of pumping is required. In this article methods for doing this by common pumps are to be briefly discussed. The sewage is collected either by the combined or separate system and carried by gravity to wells, and from these wells it is lifted through a pipe to the place and height required. In some cases it may be possible to carry away part of the sewage of a town by gravity, while another part from a low district may have to be lifted a few feet in order to enable it to get into the main sewer. At London, England, a large part of the ordinary sewage flow has been pumped for many years; and in many

European and American cities on the seacoast or along large rivers more or less lifting of sewage is done. At Boston, Mass., one half of the sewage is lifted 35 feet so as to discharge it into the ocean at high tide.

The sewer pipe which enters the wells is provided with a box at its ends, this box having screens to collect the coarse matter in the sewage. This matter consists largely of rags which are thrown into the house drains by careless people, together with sticks, lemon peel, and similar articles.

Another method of screening or straining is to pass the sewage through beds of coke; this is used when the sewage is to be lifted to filter beds where it is to be purified. In this case the iron screens intercept the rags and coarser matter, while the finer suspended matter and some of the dissolved impurities are absorbed by the coke. The coke must be removed from time to time whenever it becomes clogged.

As the sewage is mainly water, all the laws of hydraulics are applicable to its flow in pipes, and hence the computations are the same as for pumping water from a river to a reservoir. The force pumps also do not differ in principle from those used for water, but their cylinders are generally provided with hand-holes, which may be easily opened to clean the interior.

When the height of lift is low the centrifugal pump is an economical and efficient one. This is similar in principle to a turbine water wheel, except that in the turbine power is produced by falling water, while in the centrifugal pump the power is expended in order to raise water. A series of vanes arranged on a wheel and inclosed in a case is set in motion by the power, and this lifts the water through the suction pipe A, whence it enters the end of the case at B and is then forced up the discharge pipe C. At Chicago, Ill., centrifugal pumps have been used to lift the sewage into the old drainage canal that runs southward away from the city.

Syphons may be sometimes used to carry sewage over a

low lift by atmospheric pressure, but it is necessary for their successful action that a pump should be at hand in order to start the flow when it becomes interrupted. The Archimedean

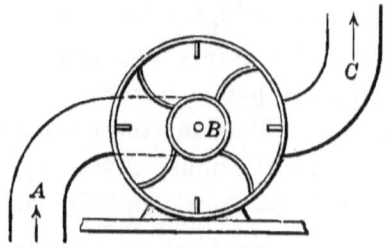

CENTRIFUGAL PUMP.

screw is another device that may be used by a low lift, but the operation of this by steam power is not economical.

The expense of the installation and operation of a pump for lifting sewage is, of course, a material addition to that of its removal by gravity, but it is never as great as that of pumping the water supply, since the lift is much lower. As a rough rule about 10 cents is the cost of lifting 1 000 000 gallons of water or sewage to a height of one foot, and on this basis the annual cost of lifting to a height of 15 feet the sewage of a town of 20 000 people, not including storm water, would be about $11 000.

58. VACUUM SYSTEMS.

The vacuum method of pumping is of course applied in every suction pump, but its extension on a large scale to the collection and removal of the sewage of a town involves some new ideas. These systems originated in Holland, where many towns are below the level of the ocean, and hence require constant pumping of some kind to remove their sewage. Originating there about 1870, the systems have spread to Belgium and France, and have been so developed that they are claimed to be efficient and economical under the conditions which there prevail.

58. VACUUM SYSTEMS.

The oldest vacuum system is that of Liernur. It employs a series of cast-iron pipes about 5 inches in diameter, from which the air is exhausted by air pumps at a central station. These vacuum pipes lead from closed basins at the street intersections to a large collecting well, out of which the sewage may run by gravity into the ocean or upon the filter beds where it is to be purified. The figure represents one of the closed basins, or evacuators, as they are called; this is made of cast iron and is about 3 feet in diameter and 3 feet high. From the houses the sewage flows by gravity through

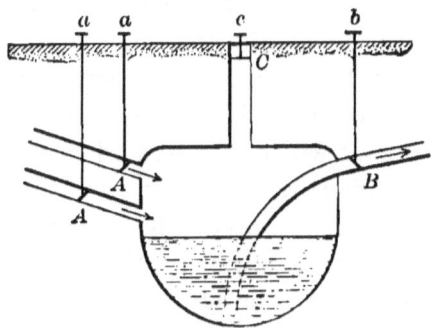

LIERNUR'S EVACUATOR.

the street pipes AA into the evacuator, from which it is to pass out through the vacuum pipe B. When the evacuator is filling, the valves in AA are open, while that in B is closed; when it is to be discharged the valves in AA are closed and those in B and C are opened, and then the atmospheric pressure drives out the sewage through the vacuum pipe to the collecting well. The evacuators are discharged one at a time by the men in their daily rounds, and during this period the air pumps keep a constant vacuum in all of the vacuum pipes.

Liernur's system was not designed to remove the liquid kitchen wastes, but merely fæcal matter with only a slight dilution of water. As the evacuator was not large enough to receive the accumulation from many houses, the street pipes held the surplus, and then by proper manipulation of the valves these were emptied one by one into the evacuator. This plan brought a suction on the house traps, and special valves were required to prevent their syphonage. In Amsterdam six of these plants were installed, each serving about 5000 people, and the sale of the excrement as a fertilizer served to materially reduce the expense of operation.

The vacuum system of Berlier is based on the same idea as that of Liernur, but the details are very different. The partial vacuum was continuously maintained in the street pipes, and the basins were made smaller and one placed in the cellar of each house. At the bottom of the basin, or evacuator, is a rubber ball attached to a float which is raised when the evacuator is sufficiently filled, and then the atmospheric pressure forces the contents through the vacuum pipes to the collecting well. It was the intention to make the system entirely automatic, but as actually built the material passes through a straining box before reaching the evacuator, and the cleaning of this box has caused considerable nuisance. Nevertheless this system has been used in certain districts of Lyons and Paris with some degree of success.

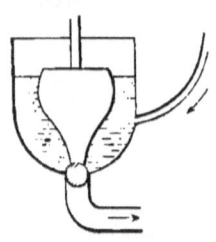

BERLIER'S EVACUATOR.

Vacuum systems have not been used in America, and since 1885 their extension has been very slight in Europe. Perhaps this is due in some measure to the circumstance that their use has been limited to excreta only, leaving the kitchen wastes to be disposed of by some other method. The Berlier system does not appear to be a more reliable method than that of Liernur, for an evacuator in the cellar must even under the best conditions be productive of some trouble. Neither system can compete in cleanliness and efficiency with that of pumping by the compressed-air method.

59. THE COMPRESSED-AIR SYSTEM.

The system of carrying away sewage by the action of compressed air was invented by Shone in 1878, and is generally called the Shone system. It has been used in a number of cities in England and in two or three in America, notably at the World's Fair held in Chicago in 1893. It consists of the combination of the separate gravity system with a method of pumping by compressed air, and has proved to be more efficient and reliable than the vacuum systems above described.

A separate system of sewers is built in the usual way over a small district and carries the sewage to a closed basin, called an ejector, in which it continues to accumulate until a valve is opened by a float. The opening of this valve allows compressed air to enter, and this drives out the sewage through a discharge pipe to the place where it is desired to deliver it.

In the following figure the level of the sewage which is shown is the lowest limit immediately after a discharge. The sewage enters through the inlet pipe AB until the float C is lifted on the spindle against the bell D, and this opens the valve E, through which the compressed air enters. The pressure of this air closes the valve B and opens F, and then the sewage is driven through the discharge pipe FG. As soon as the float falls to the level C its weight acts on the spindle so as to close the valve E; then the valve F falls by it own weight to prevent the sewage from returning to the ejector, and B is opened by the inflowing sewage.

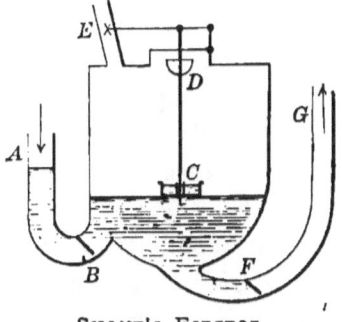

SHONE'S EJECTOR.

The largest installation of the Shone system was that at the

World's Fair in Chicago in 1893. The site was flat, and it was not desirable to turn the sewage into Lake Michigan without purification, and hence it was necessary to lift it to precipitation tanks, which was very effectively done by this method of pumping with compressed air. The separate sewage system leading from the buildings to the ejector stations was laid with clay pipes 6 and 8 inches in diameter, there being altogether 3 miles of such sewer pipe. The ejector stations were 26 in number, each having two ejectors, and there were 5 miles of cast-iron pipes from 3 to 10 inches in diameter connecting them with the central station where the air was compressed. From the ejectors to the precipitation tanks there were 4.8 miles of cast-iron discharge pipes varying in diameter from 6 to 30 inches. The lift from the lowest ejector to the top of the tanks was $67\frac{1}{2}$ feet, and the total head both static and frictional about 108 feet. The capacity of the plant was sufficient to dispose of the sewage of 600 000 people at the rate of 14 gallons per person per day.

There can be no doubt that the Shone system is a most efficient one for carrying sewage away from a flat site or lifting it to precipitation tanks. The action of the compressed air is entirely confined to the ejector and discharge pipes and it can produce no effect on the house traps. The method is hence really one for pumping alone, and its success depends upon its combination with the gravity separate system by means of the automatic ejectors. While the method is efficient, it is very costly in both construction and operation. Duplicate ejectors are necessary, so that if one be out of order the other may continue to pump, and the expenses of the air-compression plant are large. However, when sewage is to be pumped to any considerable height, and when the sewerage system of the district can be advantageously arranged for ejector stations, it is probable that the Shone method can economically compete with other methods of pumping.

60. COST AND ASSESSMENTS.

The cost of sewerage systems of different towns may be compared on the basis of the length of the sewers or on that of the population. By the first method the total cost of the system is divided by the number of linear feet of sewers, and by the second method it is divided by the population of the town. A combined system usually costs from $3 to $6 per linear foot of sewer. A separate system usually costs from $0.50 to $1 per linear foot of sewer, but to this must be added the cost of drains which are needed to dispose of the storm water. The cost of pumping and of purification works, if these are required, is to be added to the above figures.

It is not possible to say in advance that the separate system will prove more advantageous and economical for a given town. To decide this question a study of the topography and local conditions must be made by the engineer, with due regard to the experience of the town in regard to its storm water, and then comparative designs and estimates of cost will enable a decision to be made. There can be no doubt that the tendency has been decidedly toward the separate and away from the combined system since 1890, and it is hence advisable to give careful attention to estimates for carrying away the sewage and storm water by distinct and separate methods.

The expense of a sewerage system is to be met on a different plan from that of a water-supply system. Everyone expects to pay an annual fee for the use of water, but few are willing to do so in order to carry away sewage. A town can usually obtain authority to issue bonds for a water supply, because an annual income is assured which will pay the interest and probably yield a profit. The sewerage system, however, yields no income in American cities, although in Europe some cities collect an annual fee from each house connection, but it is the source of an annual loss in expenses of operation. For these reasons the progress of sewerage is slower than that of water supply, and much agitation among taxpayers is required to induce them to authorize the municipality to assume the financial burden.

The expense of the construction of the sewerage system is too great to be met by taxes in a single year, but these should be spread over several years, and hence bonds are issued to meet the outlay. The extra tax should be sufficient to meet the annual interest on the bonds, and also to establish a sinking fund which will redeem them at maturity. For example, consider the case of a town of 20 000 people whose property has the assessed valuation of $3 500 000 and whose average annual tax is 15 mills on the dollar, and suppose that a system of sewerage is to cost $60 000. If bonds are issued at 5 per cent per annum and payable in 20 years the additional annual tax must be sufficient to pay the annual interest of $3000 and also make an annual contribution to a sinking fund which at the end of 20 years will yield $60 000. If the rate of interest received in this fund is 3 per cent and it be compounded annually the annual contribution required is $2220. Accordingly the total amount to be raised by the additional tax is $5220 per annum, and this is at the rate of nearly 1.5 mills on the dollar. The annual tax rate must, therefore, be raised from 15 to 16.5 mills on the dollar for a period of 20 years in order to defray the cost of the sewerage system.

Another method is to divide the cost of construction into two parts, one to be paid by a general tax as above described, and the other to be paid by an assessment on the property along the sewer lines. This is perhaps a fairer plan, for those properties which do not front on the sewer lines cannot make connection with them and hence do not receive full benefit. The greatest benefit in all systems of drainage and sewerage is to the community at large, but usually some people receive greater advantages than others. Property along a sewered street is worth more than property on an unsewered street, other things being equal, and hence it is fair that the former should pay a larger proportion of the cost of construction. As to what this proportion should be opinions and practice greatly vary, and each town must be left to decide it for itself.

When a special assessment is laid on property along the sewer lines this should not be according to the value of the property, but in proportion to its frontage, the idea being that the advantage of being able to connect with the sewer increases the value of a property in proportion to the number of houses that can be built upon it. The assessment should be the same along branch sewers as along larger ones, and a fair plan is to make it sufficiently large to cover the cost of the smallest lateral sewers, excluding manholes, catch basins, and flush tanks. On this plan about one-fourth or one-fifth of the cost of construction is paid by the property owners along the sewer lines, and the rest is provided by an issue of bonds, for whose interest and redemption a uniform annual tax is laid on all assessed property for a number of years.

The future growth of the town is an element that must always be considered in planning a sewerage system. The records of the past give information that will be valuable in the estimation of increase in population for one or two decades, but beyond this all estimates will be merely guesses. The main sewer should perhaps be made large enough to provide for the probable increase in population for the period of twenty years, but further than this it is not wise to go. If the bonds are to be redeemed in twenty years the community will then be relieved of the taxation which these have involved and may accordingly make enlargements or extensions. In sanitary matters we look far back into the past to learn the lessons gathered from the experience of our ancestors, the present and immediate future demand of us most active work and constant vigilance, but the problems of the distant future must be left to be solved by posterity. Avoiding our mistakes and building upon what we have found advantageous, posterity shall develop more perfect sanitary regulations than now are known.

61. Exercises and Problems.

48. Collect facts regarding the sewers of Paris, their early history, and the systems now in use.

49 (*a*) Consult Gerhard's House Drainage and Sanitary Plumbing (New York, 1894), and describe the improved bell trap, the sanitas trap, and one of the traps having a mercury seal.

49 (*b*) Inspect the plumbing in a large boarding house or hotel, criticise any defects, and praise its good points.

49 (*c*) Consult Philbrick's American Sanitary Engineering (New York, 1881), and make a sketch of the old pan closet and also one of a grease tank.

50 (*a*) Consult Plumbing and House-drainage Problems (New York, 1892), and explain a few of the common defects in bath-room work which are due to carelessness of plumbers.

50 (*b*) Read an article by Corser in Engineering News, Sept. 19, 1891, and state some ideas on house sewerage from an architect's point of view.

51. Consult Hering's article in Transactions of American Society of Civil Engineers, 1881, pp. 361–386, and give an abstract of his comparisons of the combined and separate systems.

52 (*a*) Consult Adams' Sewers and Drains (New York, 1880), and make sketches showing arrangement of manholes and tidal outlets.

52 (*b*) Consult Baumeister's Cleaning and Sewerage of Cities (New York, 1894), and make sketches showing European styles of street basins.

53 (*a*) Consult Staley and Pierson's Separate System of Sewerage (New York, 1890), and make sketches of a fresh-air inlet.

53 (*b*) Describe the action of one of the mechanical flush tanks.

53 (*c*) Consult Odell's article on the Memphis sewers in Transactions of American Society of Civil Engineers for 1891, and give a description of their construction and operation.

54 (*a*) Determine the size of a circular sewer to carry off one

inch of rainfall per hour on 700 acres when the fall of the sewer is 2 inches in 100 feet.

54 (*b*) Determine the size of a circular sewer of the separate system to carry the sewage of 25 000 people when the fall is 2 feet in one mile.

55 (*a*) Consult Waring's Sewerage and Land Drainage (New York, 1889), and give a description of the sewerage system at Saratoga Springs, N. Y.

55 (*b*) Consult Baumeister's Cleaning and Sewerage of Cities and ascertain the cost of concrete sewers in European cities.

56. Consult Latham's Sanitary Engineering (London, 1875), and give an account of methods for ventilating sewers by windmills and by steam jets.

57. A centrifugal pump is to lift 150 000 gallons of sewage per hour through a height of 12 feet, the suction and discharge pipes being one foot in diameter and 172 feet long. If the efficiency of the pump is 90 per cent, estimate the horse-power required.

58. Consult Grey's report of 1884 on the sewerage of Providence, R. I., and give a fuller description of the operation of the systems of Liernur and Berlier.

59. Consult Transactions of American Society of Civil Engineers for December, 1892, and give further details concerning the Shone system in American cities.

60. A town, whose property has the assessed valuation of $7 325 000 and whose annual tax rate is 19 mills on $1, issues bonds to the amount of $225 000 to build a sewerage system, these being at $4\frac{1}{2}$ per cent interest and maturing in 15 years. An extra tax is to be laid to meet this interest and to provide a sinking fund to redeem the bonds, the rate of interest in the sinking fund being $3\frac{1}{2}$ per cent compounded annually. How many mills must be added to the tax rate for this purpose?

61. What is the distinction between a deodorizer and a disinfectant? What is a germicide? What is salt glazing and slip glazing in earthenware pipes? What sanitary work was done by Hercules under contract with King Augeus?

Chapter V.

DISPOSAL OF GARBAGE AND SEWAGE.

62. Street Cleaning.

The refuse that accumulates in the streets consists of manure, paper, leaves, and soil which has been ground into dust. The amount of dust depends upon the character of the street pavements and may be ten times as great on a macadam as on an asphalt pavement, while the amount of the other matters depends upon the character of the business and traffic. As a rough average about 1000 cubic yards of refuse per year accumulates on each mile of pavement in a densely populated city and the removal of this costs about $500 per year.

Each owner of property along a street is expected to sweep and wash his sidewalk; in towns where gutters are laid between the sidewalk and street the property owner is also generally expected to build and clean them. Generally this refuse is merely swept upon the street pavement proper, from which it is afterwards removed by the municipal authorities. Under the best regulations the property owner has nothing to do with the construction or maintenance of gutters, and indeed there should be no gutters other than those formed by the slight transverse slope of the pavement to the curbs.

The character of a street pavement influences to a certain degree the health of the adjacent neighborhood. The old cobblestone pavement, retaining foul animal matter between and under the stones, was a continual menace to health. A wooden pavement is clean when new, but after a few years of use it begins to rot and to absorb the liquid animal wastes, so that the organic dust arising from it is filled with bacteria. The macadam pavement, though excellent in suburban localities, wears quickly into mud and dust under the traffic of a city street. The Belgian pavement and the granite-block pavement, when laid with close joints on a concrete foundation, produce little dust and can be kept in a cleanly condition. An asphalt pavement is in all respects the best on hygienic grounds, as it absorbs no filth and can be cleaned with less expense than any other kind.

In villages the streets are left to be cleaned by the storm water and their surfaces are renewed by annual repairs. When the village becomes a town and builds macadam pavements an annual or semi-annual scraping is instituted to remove the dust. When the town becomes a city and the macadam pavement is replaced by stone or asphalt sweeping and cleaning must be done weekly or oftener, and when the city is a large one with heavy traffic in its streets these operations are carried on every day.

Scraping may be done on an asphalt pavement or on a tolerably smooth stone surface; when cleaning is done by scrapers it should be preceded by sprinkling, so that the dirt may be more easily removed. Scraping is most commonly used when the accumulation of material is large, as may be the case with a weekly cleaning. The scraping is done by machines drawn by men or horses, the work being begun along the middle of the street, and the material gradually moved toward the sides, where it is made into piles ready for loading into carts.

Sweeping is a better method than scraping when the work is carried on daily. Rotary sweepers having a series of brooms on a revolving axle are extensively used; when in motion the axle is inclined toward the side of the street so as to carry the refuse in that direction. Hand sweeping is also widely done to supplement the work of the machines, especially upon stone pavements. One man can clean from 500 to 1000 square yards of surface per hour, but a machine operated by a man and a horse will go over ten times as much area.

The work of street cleaning begins in the evening and continues through the night, the carting being done after midnight. In some European cities the sweeping is followed by washing; hydrant streams are turned on and men with brooms thoroughly wash the asphalt pavements. The streets in European cities are as a rule far cleaner than those in America, but excellent results have been secured in New York by Waring and undoubtedly the twentieth century will find the reform extended to all American cities. On the other hand American villages have their streets in far better sanitary condition than European ones.

The disposal of street refuse is effected by carting it into the suburbs, where it can be deposited to fill up swamps or low lands. In large cities on the seacoast it is sometimes loaded upon scows which dump it into the ocean. When the street surfaces are free from dust the sweepings are almost entirely animal and vegetable matter, so that they may be combined with the house garbage and be burned or digested in garbage furnaces. In some European localities farmers will remove this matter for use as fertilizer, but in America it rarely has sufficient value to pay for cartage. In cities which have the combined system of sewerage some of the street sweepings have been dumped into the catch basins during times of storms, but this is not a good practice, and should not be allowed except under the strict supervision of the engineer in charge of the sewers.

The problem of economical street cleaning is one of effective organization of men and methods, and hence comes under the province of the engineer. That it has not been done well and economically in American cities is mainly due to the circumstance that its direction has been intrusted to councilmen and their political adherents, instead of putting it under the charge of the city engineer. The work of the engineer, like that of the army and navy, has nothing to do with any political party, but is conducted for the welfare of the community only. To secure the highest efficiency and economy of street cleaning and other public engineering works the same methods must be adopted as those used by a private corporation, namely, to select the best men, let out the work to contract when advisable, and by vigilant superintendence and inspection secure the maximum performance with the least expense.

63. REMOVAL OF GARBAGE.

The solid refuse of kitchens is of two kinds, the inorganic and the organic: the first consists of dust, ashes, and pieces of metal; the second consists of animal and vegetable matter and is called garbage. In the village the disposal of these is left to the householder; he throws the ashes upon the fields or uses it to build sidewalks, and he utilizes the garbage as food for animals. When the village becomes a town the garbage is often carried away by farmers, and when this proves unprofitable a public scavenger is appointed to collect it and cart it into the country to be dumped upon waste fields. As the town grows into a city the method of collection is more systematized, and the garbage may be burned in special furnaces in order to prevent the evils due to its decomposition.

The separation of the ashes from the garbage should be insisted upon in all systems of public removal. In the town each householder disposes of his ashes, while the municipality removes the garbage. In the large city both are to be removed by the authorities, but in separate vessels. The mineral refuse often has a market value in Europe; pieces of metal are sold to junk dealers, ashes are used to make slag cement, and cinders are worth something for fuel; in some American cities the time has already come when contractors will remove the mineral refuse in order to sort it over in the suburbs. The value of the garbage is usually less, but in Europe it is sometimes sorted; the rags are worth a little, and the meat bones are utilized as a fertilizer. Generally, however, the garbage of American cities is without value, and the problem to be solved is how to remove and dispose of it with the least expense.

An imperfect method of garbage removal, often used in towns, is to require each householder to provide his own vessel, the contents of which are dumped into an open wagon in its weekly rounds. This is objectionable because the dirty vessel is not always cleaned by the servants, and because the garbage on the wagon causes more or less unpleasantness in the streets. The ideal method is to have the municipality furnish and clean the vessels; starting on his round with a wagon load of clean empty vessels, the scavenger leaves one at each house and takes a full one in its place. These vessels should have covers which can be fastened with a hasp, and thus the transportation through the streets causes no offense. The size of the vessels is such that they may be filled during the interval between two collections, and under the best regulations this interval is not longer than two days.

When garbage is dumped upon scows to be carried out to sea or upon suburban fields to serve as manure it is sometimes sprinkled with lime in order to neutralize the odors of decay. At this stage the street sweepings may be combined with the garbage. The method of disposal at sea is a satisfactory one if the scows go several miles away from land, but otherwise the garbage may be washed back upon shore by the currents and storms. The method of disposal upon fields is rarely an efficient one, unless it be carried far into the country and only a small amount applied in one place. Thus to render either method fully satisfactory a large expense for transportation results.

The theory of the purification of garbage by dumping it into water or by spreading it on land is the same as that given in the first chapter to account for the transformation of dead into living organic matter. In both cases oxygen (O) is furnished to attack the carbon (C) and thus carbon dioxide (CO_2) results. Next nitrogen (N) and hydrogen (H) are liberated and ammonia (NH_3) is formed, which upon further oxidation becomes nitrous acid (HNO_2) and nitric acid (HNO_3); these acids by combination with metallic compounds produce nitrites (MNO_2) and nitrates (MNO_3). Thus under the favorable condition of the presence of abundant oxygen the dead organic matter becomes resolved into harmless gases and solids. But if sufficient oxygen be not furnished the process of decomposition becomes more complex and results in putrefaction, whereby bad-smelling gases are evolved. This occurs by the combination of the carbon dioxide with the ammonia and other substances to produce gases which not only cause much offense, but are popularly supposed to be more injurious to health than the products of decay under common conditions.

In all methods of the disposal of garbage the aim should be to remove it from the houses at frequent intervals and before the process of decay has fairly begun, and to deposit it under conditions where oxygen may have opportunity to attack every part so that putrefactive decomposition may not take place. This is readily done at an isolated country house, but in a city, where the refuse of thousands of houses is constantly accumulating, it is not an easy matter. Hence other methods for getting rid of garbage have come into use, namely, methods of destruction by fire and by heat, and these will now be briefly described.

64. Cremation of Garbage.

The method of disposing of garbage by burning it in the kitchen fire is one that has long been practiced, but furnaces for the cremation of the garbage of a town did not come into use until after 1880. Of all the methods for destroying decaying substances that by fire is undoubtedly the most effective, as thus all accompanying bacteria are killed, and the organic matter is completely oxidized into gases and only ashes are left behind. The objections to the method are two: it is expensive, and it is liable to produce offensive odors. Both of these objections have been gradually diminished by the experience gained, so that at present a considerable number of cities in Europe and America are burning their garbage and street sweepings with economy and success.

In the cremation of garbage special furnaces lined with fire brick are employed, and a cross-section of one form is shown in the following figure. A represents the furnace chamber,

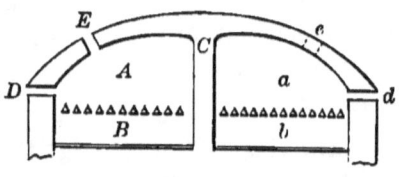

SECTION OF GARBAGE FURNACE.

B the ash pit below the grate, E one of the openings through which the garbage is dumped, and D one of the stoke holes through which the burning matter may be stirred. The width of the chamber may be about 5 feet and its length about 15 or 20 feet. The fuel used is petroleum, which is injected through pipes at a number of places both above and below the grate. The gaseous products of the combustion may pass to the end of the chamber A and thence into a, where they are still further consumed by burning oil. The disposition of the gases may be directly into the atmosphere by means of a tall chimney, or they may be carried into the fire under a steam boiler in order to be more completely oxidized.

Another and simpler arrangement is to use both the furnaces A and a for burning the garbage, the gases passing up a common chimney. Natural gas and gas made from bituminous coal have been used for fuel instead of petroleum. In the early furnaces coal was used, but this requires very careful stoking in order to maintain the high degree of heat necessary for effective combustion. The above figure gives only a general idea of a crematory, as the details of arrangement are quite different in the furnaces of different patentees.

The process of cremation is widely used in England and to a less degree on the European continent. It was first employed in America in 1885 to burn the garbage at army posts, but has since spread widely and has been used in probably 20 or 30 cities. The expense of burning one ton of garbage may be roughly estimated at 75 cents, and, since the amount of garbage produced in a city is from one-third to two-thirds of a ton per person per year, the probable annual cost of disposal in this manner ranges from 25 to 50 cents per person. Efforts have been made, especially in England, to utilize the escaping hot gases for heating the steam boilers at electric power stations, but it does not appear that this has gone further than the experimental stage. Attempts have also been made to utilize the ashes in making mortar, but no information is at hand regarding the success which has been attained.

One advantage of the cremation method is that all kinds of garbage and street sweepings, even when mixed with ashes, may be simultaneously burned, and that dead animals as large as horses may be cremated in the largest furnaces. The disadvantage is in the expense and in the complaints of odors from the escaping gases. Notwithstanding the somewhat wide use of the method since 1890, the tendency seems to be to consider the objections as valid ones and to regard the method of disposal by digestion as a more economical and cleanly one.

65. Digestion of Garbage.

The method of disposing of garbage by digestion consists in heating it in closed vessels until it becomes separated into water, grease, and solid nitrogenous matter. It is important that the garbage should be free from ashes and metallic substances, and that it should not contain the dead bodies of large animals. When all these things are collected indiscriminately the garbage proper must be sorted out before it is put into the digestors. The word reduction is frequently used instead of digestion in designating this process.

A digestor is a vertical steel tank which can be tightly closed by a cover after it is filled with garbage. To this tank steam pipes are attached, through which steam is forced for several hours and thus the garbage is cooked or digested. This cooking produces water, which falls to the bottom of the tank, and lighter nitrogenous matter, which tends to rise, while grease is mixed with both of these.

The arrangements of the different digestors in use differ in many details. One of the best forms is about 6 feet in diameter and 20 feet high, the upper 15 feet being cylindrical, and the lower 5 feet being conical, with a valve at the bottom for drawing off the liquid. The garbage is dumped into the top of the digestor and rests upon a grating at the bottom of the cylindrical part. The steam is then applied for six hours, and during this process the waste gases and steam are carried out in a pipe at the top of the digestor and condensed in a water tank. The liquid matter is then drawn out at the bottom of the cone and collected in vats, where the grease rises to the surface and is skimmed off, while the remaining dark-colored liquid is run into the sewers. The more solid matter, called tankage, is taken out at a door just above the grating, put under a press in order to expel the water and grease still remaining, dried in ovens, and then ground into powder. The grease is sold to soap manufacturers, and the pulverized tankage is sold to farmers for a fertilizer.

While the expense of construction and operation of a digestion plant is greater than that of a crematory, the market value of the grease and fertilizer is such that the net expense of disposing of garbage by digestion appears to be less than that by cremation, especially when the quantity is large. The system has been used in Buffalo, Boston, Philadelphia, and other cities with satisfactory results, and it is found that the odors from the operation are in general not offensive. The methods for the collection of ashes and garbage are rapidly being systematized in large cities, so the separation of the two kinds of refuse does not need to be done upon a common dumping ground, and as these become more and more perfect the ashes and mineral matter will have a higher value, and the garbage digestion can be done more economically. Paper is not desirable in a garbage digestor, and hence it should be put with the ashes or be collected separately.

From a purely sanitary point of view almost the only objection to the digestion process is the dark-colored liquid, which is often run off into the streams. This contains much organic matter, and although the bacteria have been killed by the cooking and boiling others will soon be supplied from the water of the stream, and decay will then take place. The remedy for this is to forbid the introduction of such liquids into rivers, and to require them to be purified either by the method of filtration or by that of irrigation described in Arts. 72 and 73. When this dark-colored liquor can be turned into a sewer the objection loses its force, for the volume is very small compared with that of the sewage itself.

66. Cleaning of Cesspools.

A privy and a cesspool are essentially the same in principle, both being holes or vaults in the ground, but the former is shallow and open at the top, while the latter is deeper and has its top covered. The common country privy is usually an offensive place, but a cesspool may be arranged so as to be cleanly in comparison. The privy receives only human excrements, but the cesspool usually receives both these and the kitchen drainage, and often also some of the roof water. Wherever population is dense the privy system is sure to produce disease; for instance, the epidemics of typhoid fever that prevailed in 1898 at military camps in the United States and West Indies were probably caused by common flies, these transferring the germs of the disease from the privy deposits to the food of the soldiers.

The hole which forms the privy or cesspool is walled up with stone, and a common practice in the country is to have the walls of loose stone, so that the liquid material may leak through them into the surrounding soil. The result of this is a gradually increasing pollution of the soil, and often a neighboring spring or well becomes contaminated so as to cause disease. For instance, near Easton, Pa., in August, 1898, twelve cases of typhoid fever and two deaths resulted from the use of spring water which had become infected from a cesspool; the spring had previously been condemned by the board of health, but in spite of the warning the people of the neighborhood continued to drink its water. Thousands of such cases are given in medical journals and in reports of boards of health.

In a well-regulated village the privy and cesspool vaults are required to be built with tight walls and bottoms, so that no leakage into the soil may occur, and their contents are to be removed at regular intervals under the supervision of the board of health. This operation is done by a contractor who has the special apparatus for effecting the removal with the least nuisance, and who is required to use disinfectants upon the material, the vault, and the apparatus in order to destroy the odors, kill the bacteria, and thus prevent contamination of the air.

Earth, charcoal, ashes, and similar substances are deodorizers, but not disinfectants. A substance is said to be a disinfectant when it acts upon decaying matter so as to stop the process of decay, and this is done by its poisonous action upon the bacteria. When the bacteria are deprived of life the decomposition ceases and the gases which accompany decay or putrefaction are no longer evolved. Carbolic acid, iron sulphate, chloride of lime, sodium hypochlorite, and many other chemicals are efficient disinfectants whose proper application in the cleaning of a privy or cesspool will prevent all offense and render the operation harmless. The men who do such cleaning cannot, however, be trusted to effect thorough disinfection, and it is hence important that an inspector of the board of health should always be present to strictly enforce the regulations.

The pneumatic cart is the best apparatus for cleaning a privy or cesspool vault. This has an air-tight cylinder mounted on wheels and in general appearance resembles a watering cart. The material in the vault should be in a semi-liquid state, which can be effected by the addition of water, if necessary. A hose leads from the cylinder to the vault, and the valve in the hose is closed until the air has been exhausted from the cylinder by means of an air pump; the valve is then opened and the atmospheric pressure forces the liquid up into the cylinder. This process is repeated until the cart is filled or the vault entirely emptied. Almost the only danger in this operation is the contamination of the atmosphere by the air pumped out of the cylinder. When the pump is driven by a portable steam engine the exhausted air may be pumped through the fire under the boiler; when the pump is driven by hand the exhausted air may be carried into a barrel of water containing carbolic acid or some other powerful disinfectant in solution.

The disposal of the material is the hardest part of the problem. In Europe, where manure is valuable, it may often be sold to the peasants; in America farmers will allow it to be spread on their fields, but will rarely pay for it. In any event the method of disposal as manure is restricted to the country, and a city which uses the cesspool method is forced to dump the material into streams or to bury it in the ground. The nuisance, vexation and expense of the cesspool method become in time so great that the town abandons it and substitutes a water-carriage system of sewage removal.

The privy and cesspool methods must long continue to be used in the country and in villages, but the fact that diphtheria and typhoid fever are more common in the country than in the city should serve as a continual warning. An ounce of prevention is worth more than a pound of cure in all sanitary matters. The daily use of dry earth in country privies, municipal regulations for tight vaults and proper removal in villages, the organization of an efficient board of health in towns and frequent sanitary inspections by it, are preventatives which are too often regarded as unnecessary because the health of the community appears to be fair.

67. Sewage and its Decomposition.

Sewage is water containing the decaying matter of household wastes in suspension and solution, and, as stated in the last chapter, the weight of total solids is always much less than one per cent of the whole. European sewage has the total solids from 1000 to 3000 parts per million, but in American sewage, owing to the larger volume of water supply, the proportion is generally less than 1000 and often as low as

500 parts per million. Of the solid matter about two-thirds is organic, and the object of the sewerage system is to remove the sewage from the town before the decomposition of this organic matter has fairly begun.

As the water supply flows from the faucets into the house fixtures air is mingled with it, so that the sewage in its flow through the house drains contains much dissolved oxygen. This oxygen immediately attacks the organic matter and the bacteria begin their useful work of decomposition and increase in number and activity. As the oxygen becomes used up in this work the bacteria decrease in number and the decay finally goes on by the method of putrefaction which was explained in Art. 63. Sewage, therefore, presents different characteristics at different stages of its history.

Fresh sewage is that flowing from the house drains into the sewers; its odor is not bad, as the decomposition is not fairly under way. Stale sewage is that about one day old or at least of such an age that the odor is unpleasant. Septic sewage is so old that the decomposition is going on by the putrefactive method and very ill-smelling gases are evolved. In stale sewage the bacteria have reached their highest development, both in respect to number and activity, while in septic sewage the number is comparatively low.

The following analyses of the Massachusetts State Board of Health, as stated by Clark in 1898, give an idea of the changes which occur in sewage from the fresh to the septic state, the figures being in parts per million:

	Fresh.	Stale.	Septic.
Free ammonia,	26.0	45.0	55.0
Albuminoid ammonia,	11.8	10.5	5.5
Nitrogen as nitrites,	0.21	0.0	0.0
Nitrogen as nitrates,	1.00	0.0	0.0
Oxygen consumed,	85.0	48.0	25.0
Bacteria per cubic centimeter,	1 950 000	3 800 000	500 000

Here it is seen that the free ammonia increases with the age of the sewage, while the albuminoid ammonia and the amount of oxygen consumed decrease. In septic sewage, therefore, a considerable proportion of the organic matter has become transformed into carbon dioxide, ammonia, and other gases, but no nitrification has occurred, and in fact the nitrites and nitrates of the fresh sewage appear to have been resolved back into ammonia.

When sewage is discharged into a river this is usually done during the stale state or during an early stage of that state. The river water then furnishes an additional supply of oxygen, so that decomposition goes on by means of nitrification and the septic state is never reached. So also when sewage is purified by filtration through earth the septic state is not reached. Indeed, the subject of septic sewage is one concerning which little is known, for the almost universal practice has been to dispose of sewage as quickly as possible and before putrefaction can take place. Judging from the number of bacteria, stale sewage is more dangerous than septic, but if the warning given by the odor is of any value the septic state is one that should be avoided.

The introduction of the acid wastes of factories into the sewers is not regarded as desirable, because these acids are poisons which kill the useful bacteria of fresh sewage, and hence the decomposition may proceed by the septic method with the result of giving much offense by the foul gases. Such acids, however, are usually admitted to the sewers, and as a consequence the sewage of manufacturing towns has a very complex composition and is more liable to cause river pollution than the sewage from houses.

The methods for the disposal of sewage are now to be taken up and discussed. Its disposal in streams is first to be noted, and then chemical precipitation, filtration through sand-beds, and utilization on farms will follow. In all these methods the end to be constantly kept in view is to resolve the organic matter into harmless constituents in such a manner that neither air nor water may suffer pollution.

68. Disposal of Sewage in Rivers.

When a brook or a river flows through a village nothing seems more natural to the inhabitants than to throw garbage and refuse into it. Privies are often built overhanging it, the waste of kitchens is turned into it, and rubbish of all kinds is dumped upon its banks. When the stream is a mighty river the matter thrown into it by a few men is of little moment, as it is quickly diluted and oxidized by the great mass of water; but when the stream is a very small one it becomes fouled so as to be more uncleanly than a sewer. The village itself may not feel the effect of this pollution, but the next village lower down on the stream may be seriously affected by the impure water.

An impure stream becomes purified by a flow of sufficient length if additional organic matter is kept out of it. Aeration and sedimentation constantly go on, and by these processes impure water may become pure. Hence it has been said by some that a flow of ten miles renders sewage harmless in a river, while an English commission maintained that there is no river in England or Scotland long enough to effect the oxidation of sewage. It is recognized by all, however, that a small amount of sewage in a large river will soon be purified so as not to be injurious, and that a large amount of sewage in a small river cannot be purified. Between these two extremes there has been found every variety of opinion and practice.

It has been shown by Mason that in 1890 germs of typhoid fever were carried 26 miles by the Mohawk and Hudson rivers, and Sedgwick has also shown that in 1892 such germs were carried 25 miles by the Merrimac River. For these distances at least the flow of the river did not effect purification.

Negative evidence indicating that no disease was known to have been caused by sewage after a flow of four or five miles in a river has little value in comparison with a few positive facts.

The amount and character of the sewage, the volume and velocity of flow of the river, and the kind of fishes and vegetation found in it are three controlling factors in the question of river pollution. If the volume of sewage be larger than one-twentieth of the river water, or if it contain acids of manufactories, the conditions are very unfavorable for its purification. If the velocity of the stream be small and there be no fishes or vegetation in it the conditions are also unfavorable. Fishes eat the grosser particles of organic matter, and vegetable growths absorb the dissolved matter, but the acid wastes of factories may kill both fishes and vegetables as well as the bacteria of the sewage and hence retard the purification. When a river is sluggish the available oxygen is soon used up by the sewage and the septic state of putrefaction may ensue. Leeds reports that in 1883, when the Schuylkill River near Philadelphia was frozen over, the gaseous products of this putrefaction were so abundant that they escaped through holes in the ice, and when set on fire rose a foot in height.

By far the larger number of American cities turn their sewage into rivers or into the ocean. The disposal at sea is comparatively unobjectionable if the discharge be made at such times that it will not be returned to the shore by the tides, but this cannot always be secured. The disposal by discharge into a river is gradually becoming regarded as a menace, if not an actual danger, to towns which take their water supply at points below, and laws have been enacted in some states to prevent the contamination of rivers. These laws are necessarily general in their wording, and the specific details are left to be supplied in each case by the boards of health. For instance, the law may forbid any person to put "any polluting matter into a stream used as a source of water supply," but it must be left to the board of health to determine whether the sewage of a village is or is not polluting matter.

As a consequence of these laws many towns and cities have been, or will soon be, obliged to install precipitation or filtering plants for the purification of their sewage. These plants do not render the sewage so pure that it is safe for drinking, but they can generally make it as free from bacteria as the water of the river into which it is to be discharged. The degree of purity required by the boards of health will depend upon the relative volumes of the sewage and river flow and the distance to the next town which takes its water supply from the river. The criterion of the number of bacteria per cubic centimeter is often adopted, and an additional one is that of the organic matter remaining after purification. If the river water contains 500 bacteria per cubic centimeter and the purified sewage only 400, and if the amount of organic matter in the river water is about the same in both, it would seem that the artificial purification has been carried as far as practically necessary. These requirements and the methods for enforcing them are as yet in the early stages of development, but the guiding principles have been well established, and the time is near at hand when they will certainly become a powerful factor in the progress of sanitary science.

69. Screening and Aeration.

A screen or sieve may be used to effect a slight degree of purification in sewage by intercepting a part of the suspended organic matter. By passing the sewage through several sieves, each finer than the preceding one, a very large proportion of the suspended matter may be removed and the sewage be rendered much clearer in appearance. A sieve, however, does not remove the dissolved organic matter, and this is usually the part which is in the state of most active decay and hence the most dangerous.

Sewage must be screened to a certain degree whenever it is pumped or whenever it is distributed over filter beds. A single box screen at the end of the sewer will catch the rags and sticks and thus render the sewage clear enough to be pumped, but for distribution on filter beds a series of screens should be used. For this purpose the sewage may be led into an open masonry channel of rectangular section in which vertical screens or sieves are placed at intervals; if the velocity of flow be small some sedimentation will also occur. After a day or two the matter accumulated at the screens may be removed; this may be digested or burned like garbage, or if such methods are not available it may be mixed with sawdust and then be spread upon the fields to serve as manure.

Screening through coke is a process which has received much attention. Coke is the porous charcoal obtained by heating bituminous coal in the absence of air. When sewage is passed through it the suspended matter is strained out, but this accumulates in the coke and after a week or more a new supply is needed. The coke is spread in a bed about a foot thick supported on wire screens and the sewage is passed through it by the action of gravity. The effect of the screening will vary with the kind of coke and the rate of passage through it, but a large part of the suspended organic matter can be generally removed. When the coke becomes clogged it is removed and burned under boilers, but this often gives offensive odors, and hence the plan of first heating it in an oven to extract the grease has been tried.

When a bed of coke is thick and the rate of passage of the sewage is slow the process becomes filtration instead of screening, for thus the bacteria are given time to do their useful work on the dissolved organic matter. Ashes, cinders, peat, and similar substances have been used instead of coke for such beds, and usually the action which results is a combination of screening and filtration. The process of screening alone may be carried on continuously, but the process of filtration should be intermittent in order that air may enter the beds and furnish oxygen to the bacteria. In screening processes the bacteria are decreased in number merely because the amount of decaying organic matter is decreased, but in true filtration processes the bacteria are almost entirely removed because their work is done, the organic matter having been totally decomposed. In all screening methods it is desirable that the sewage should be as fresh as possible in order to avoid the offense which might be caused by foul organic matter, but it will be seen in Art. 70, where the subject of filtration is discussed, that it is desirable that the sewage should have reached the stale stage before it is applied upon the beds.

Sedimentation is a slow process for the purification of water, and it is entirely impracticable for sewage unless the precipitation be accelerated by chemical methods. Sewage, even after straining, contains so high a proportion of organic matter that any sedimentation occurring by the action of gravity alone could effect only a slight degree of purification. Moreover, sedimentation requires so much time that the decomposition would probably become septic and thus evolve odors which would cause great offense. The storage of sewage for purposes of sedimentation is hence wholly impracticable. Storage for the purpose of changing solid matter into gases through sceptic decomposition is another question.

The aeration of sewage consists in supplying it with air, so that oxygen may be furnished to the bacteria and enable them to continue the useful work of purification. Aeration is always beneficial in removing odors, and effective purification will result if sufficient air is furnished during a considerable period of time. It must not be forgotten that bacteria require both time and oxygen, and hence it is plain that one violent aeration of short duration will not be as effective as a number of aerations of less intensity with suitable intervals between them. In all methods of aeration the sewage must first be screened and strained in order that the effect of the oxygen may be concentrated upon the organic matter in solution.

Combined screening and aeration may be done for a small volume of sewage by the use of a number of horizontal sieves, one placed above the other and the finest being at the bottom. As the sewage falls through the intervals between the sieves it becomes aerated more and more, and after passing the last sieve the suspended matter is in large part removed and the bacteria are actively at work on the dissolved matter. This method generally requires the sewage to be pumped, and the expense of this, together with the fact that the purification is not as complete as by other processes, has prevented it from coming into use much further than the experimental stage. Artificial aeration by agitating the sewage with revolving wheels and by blowing air into it have also been tried, but such methods are also expensive and can be used only on a small scale.

It is seen by the above discussion that screening and aeration either separately or combined do not constitute a method for the effective purification of sewage except in special cases where the volume to be treated is small. They are, however, valuable adjuncts in the three systems of purification which are to be described in the following articles. These three systems of sewage disposal are called chemical precipitation, intermittent filtration, and broad irrigation, and it has been demonstrated that by the second and third of these systems sewage may be completely purified.

70. Chemical Precipitation.

The theory of the precipitation of sewage by means of chemicals is the same as that given in Art. 25 for water. The chemical substances are added in the form of a solution, and their reaction results in a precipitate which falls by the action of gravity and drags down with it the suspended organic matter of the sewage.

Lime, or calcium monoxide (CaO), when placed in water forms calcium hydrate (CaO, H_2O), and this solution when added to sewage containing carbon dioxide (CO_2) results in the formation of calcium carbonate ($CaCO_3$), which is precipitated. The amount of lime added to the sewage averages about 200 parts per million, or about 1600 pounds of lime to 1 000 000 gallons of sewage. As decaying organic matter gives off carbon dioxide, this is always found in sewage, and sufficient lime should be added to completely absorb it, if the best results are to be obtained.

Alum, an aluminum-potassium sulphate, is a precipitant whose action is much quicker than that of lime, the substance thrown down being aluminum hydrate, according to the reaction given in Art. 25. The quantity of alum required is only about one-half of that of lime, but its cost is about three times as great. Accordingly a mixture of lime and alum is frequently employed in the proportion of about one part of the former to four parts of the latter.

Copperas, or ferrous sulphate ($FeO, SO_3, 7H_2O$), and ferric sulphate ($Fe_2O_3, 3SO_3$), are also used for clarifying sewage, the latter in particular forming a ferric hydroxide ($Fe(OH)_3$) which is very quickly precipitated. Copperas generally requires the addition of lime in order to make it efficient, and the precipitated substance is a ferrous hydrate ($Fe(OH)_2$) or a ferrous carbonate ($FeCO_3$). It cannot be said, however, that either of these precipitants is always more efficient than alum or lime, for it is found that different kinds of sewage require different treatment, and in each case some experimenting is necessary to determine the most advantageous and economical chemicals.

By chemical treatment all of the suspended organic matter and about one-fourth of the dissolved organic matter may be removed from sewage. The matter precipitated in the bottom of the tanks is called sludge, while the purified sewage is generally termed the effluent. The action of the chemicals also greatly reduces the number of bacteria, those not precipitated in the sludge being poisoned, so that under very careful treatment the effluent may contain less than five per cent of the bacteria of the original sewage. Hence the effluent is so far purified that in many cases it can be turned into rivers without fear of causing pollution.

A precipitating basin is often formed by an open conduit with a very slight slope which is built of concrete and has gates to regulate the flow. In the figure the plan of such a conduit is shown, its width being about 5 feet, its depth 3 feet, and its length 200 feet, the capacity being sufficient to treat the sewage of about 1000 people. The sewage enters at A, either from the sewer or from a pump which has raised

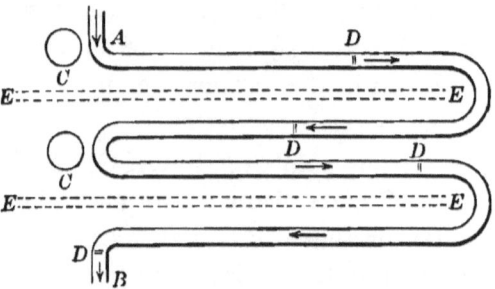

PRECIPITATING CONDUIT.

it to the required elevation, and passes out at B. The chemicals are in solution in the tanks C, from which they flow directly into the conduit, and the gates D regulate the flow. EE are tracks where a wagon runs when carrying away the sludge, this operation being done about twice a week. A duplicate conduit is usually provided, so that when one is being cleaned the other may be in operation.

The intermittent method of operating a precipitating basin is to fill it with sewage, then to add the chemicals, and after the contents have stood for a few hours to discharge the liquid portion, the sludge remaining during several such fillings. The continuous method consists in keeping the basin nearly full and allowing the effluent to flow out as fast as the crude sewage enters, the chemicals being continually added. Experience indicates that the continuous method is the better one, but when conduits are used much care is required in both methods that the final discharge be not made so rapidly as to cause the soft sludge to mingle again with the effluent.

Vertical tanks are used more extensively than conduit basins when a large volume of sewage is to be treated, as they occupy less space and the sludge is spread over a smaller area. Such a tank may be a masonry basin when the ground is low below the main sewer outlet so that an excavation can be made, or it may be made of riveted steel plates and be placed above ground when the sewage is to be pumped. In the figure the first diagram shows a masonry tank where the sewage enters by a conduit at A, whence it falls to D and

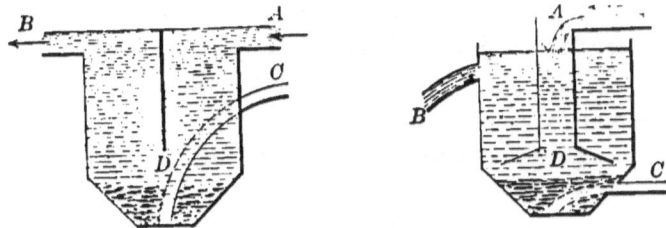

PRECIPITATING TANKS.

then rises to flow away at B; the soft sludge is removed from the bottom by means of a pump attached to the pipe C. The second diagram shows an elevated steel tank where the sewage is pumped up to A, whence it falls through the vertical pipe AD and the effluent passes out at B, while, as before, the sludge is pumped out through C. In both cases the operation is continuous, the dissolved chemicals being constantly added to the crude sewage as it enters.

The disposal of the sludge is the most difficult part of the problem, for this is merely soft mud when it is taken from the basins. In disposal works for a small town shavings, peat, or leaves are mixed with the sludge, so as to enable it to be more easily handled, and it is then burned or spread upon the fields. Sometimes it is run into scows and dumped at sea. The most advantageous way, on the whole, seems to be to run the soft sludge into an apparatus called a filter press, which squeezes out the liquid portion and forms the solid portion into cakes. The liquid which is pressed out is run back into the tanks to be again precipitated, while the cakes are buried in the ground or burned in a garbage crematory. The cakes have no value for manure, as the precipitate is in large part mineral matter from the chemicals; when the soft sludge is mixed with sawdust or compost the compound formed may have a slight value, but usually not enough to induce farmers to cart it away.

This method of sewage disposal has been much used in Europe, and since 1890 has been introduced in a number of American towns and cities, among which may be mentioned Worcester, Mass., South Orange, N. J., Echota, N. Y., and Chautauqua, N. Y. The largest installation of the system thus far in America was at the World's Fair of 1893 in Chicago, where the sewage collected by the Shone system was forced to a vertical pipe 30 inches in diameter around which four precipitating tanks were placed of the general style shown in the second diagram of the above figure. Each tank was 32 feet in diameter at the top and the height of the straight sides was 32 feet also, the combined capacity of the four tanks being 237 000 gallons. The effluent was run into Lake Michigan, while the sludge was pressed into cakes, which were burned in the garbage crematory. The cost of the method including the disposal of the sludge may be stated at about 50 cents per person per year as a rough average for American cities.

The effluent, although much clearer and purer than the

sewage, is far from being like pure water. The sewage has about 40 per cent of the organic matter in suspension and 60 per cent in solution, and if all of the former and one-fourth of the latter be removed there still remains in the effluent 45 per cent of the original organic matter. The bacteria, it is true, have been enormously reduced in number, but as soon as the effluent is run into a stream where oxygen is furnished to it they immediately multiply and the evils of decomposition result. The experiments of Hazen in 1889 have shown that under good conditions alum removed 91 per cent of the bacteria, ferric sulphate removed 95 per cent, lime removed 97 per cent, and copperas with lime added removed 98 per cent. As the average number of bacteria in a cubic centimeter of sewage is about 1 000 000, the effluents still contained from 20 000 to 90 000 bacteria per cubic centimeter, or a thousand times as many as found in good potable water. With respect to the dissolved organic matter Hazen found that ferric sulphate removed less than one-half, while lime removed less than one-fifth. On the whole it appears impossible to purify sewage by any process of chemical precipitation so completely as by the methods of intermittent filtration and broad irrigation.

71. INTERMITTENT FILTRATION.

The method of purifying sewage by filtration is founded on the same principles as those set forth in Art. 28 for the artificial filtration of water. Sewage is a very impure water, but not much more impure than the surface drainage of some pastures and swamps; by passing it through soil at a slow rate and supplying sufficient air to enable the useful bacteria to work the dead organic matter becomes completely changed into harmless gases and mineral compounds, so that the resulting effluent is clear and pure water.

In the filtration of a water supply either the continuous or the intermittent method may be used, but with sewage the intermittent method of operation seems always necessary, because its larger proportion of organic matter requires the presence of a greater amount of air. When a filter bed is drained of its liquid contents the surfaces of the sand grains remain still covered with thin films of water, and the air of the atmosphere enters around these films and thus furnishes oxygen to the bacteria which are engaged in hastening the chemical operations of oxidation and nitrification.

In water filtration the greater part of the bacteria are at work in the top of the layer of sand, but it has been the aim in constructing filter beds for sewage to extend their operation further downward. This has been done by using coarser materials, such as gravel, broken stone, and cinders, so as to prevent the clogging which results from fine sand. A portion of the sludge collects upon the surface, however, and from time to time this may be removed and cleaned, as in water filtration.

The rate of filtration of sewage must be slower than that of water, and hence a larger area is required for a given volume than in water purification. The rates for water range from 30 to 100 gallons per square foot per day, but for sewage the rate may be as low as 1 or 2 gallons per square foot per day. As a very rough rule one acre will purify the water supply for a city of 10 000 people or more, but it will scarcely be sufficient for the sewage of 1000 people.

The most favorable location for sewage filter beds is along the bank of the stream into which the effluent is to be discharged. The average size of the beds is about one acre, and each of these has its own main underdrain, into which the smaller lateral drains lead. It is generally too expensive to build masonry walls to separate the beds, and hence wide earthen embankments are made, these being thoroughly rolled. The bottoms of the beds are usually natural earth, in which the underdrains are laid in trenches. These drains are provided with valves at the places where they pass through the embankments, and it is by means of these that the rate of filtration is regulated. When the soil is of favorable quality the expense of preparation of a series of beds may be as low as $2000 per acre, so that the entire sewage-filtering field for a city, notwithstanding its larger area, may cost less than the filter beds for the water supply. A sedimentation basin is not needed, because the effluent is not to be used, but a receiving basin for the crude sewage is necessary so as to properly distribute it over the different beds.

It has been found that fresh sewage requires a slower rate of filtration than stale sewage. This is due to the circumstance that time is required to develop the full number of bacteria needed to perform the work of oxidation and nitrification. In stale sewage the bacteria have reached their maximum development and hence its purification takes place more rapidly in the filtering material than does that of stale sewage, while less sludge is also deposited on the surface. The question of the use of septic sewage on filter beds is one that has also been studied; the amount of sludge in this case is smaller, because some of it has been changed into gas, but the bad odor of the gases evolved in the storage and application of sewage in this form seems to be a serious objection to its use.

The frequency of draining and aeration will depend upon the rate of filtration and upon the degree of purity required in the effluent as well as upon the character of sewage and filtering material, and is to be determined in each case by experiment. For a very small volume of sewage a method of forced aeration has been advocated and tried; here the small beds are supported in boxes with a sieve bottom and air is blown through them by a pump after they have been drained.

The frequency of cleaning the surface depends also upon the character of the sewage and the rate of filtration, and the aim should be to render the intervals as long as possible. In general a period of two or three months elapses between the times of cleaning, but the most perfect filtration would be like that which occurs in nature, where the surface requires no cleaning. In nature this is usually effected through the absorption of the organic matter by growing plants, and the artificial application of this process to sewage leads to the method of irrigation which is to be discussed in the next article.

With proper attention sewage can be purified by intermittent filtration so as to be undistinguishable from pure water by either chemical or biological analyses. The organic matter has been transformed into carbon dioxide and ammonia, the ammonia has combined with the mineral substances in the soil to form nitrates, and the bacteria have been reduced to a number fewer than is found in natural potable waters. Sentiment, of course, forbids the use of the effluent as drinking water, but there is no scientific reason why it may not be used as such with entire safety if comprehensive analyses and their interpretations so indicate. Chlorine is not removed by filtration, and the amount of this in the effluent will hence be found very high, but this has no bad influence on health, and its presence has a wholly different interpretation from that of a similar quantity found in the natural waters of the neighborhood. The following analyses of sewage and its effluent from one of the experimental tanks of the State Board of Health of Massachusetts will give a typical idea of the results of intermittent filtration:

		Sewage.	Effluent.
Total solids, parts per million,		466.8	214.4
Inorganic matter, " "		338.0	202.7
Organic matter, " "		128.8	11.7
Chlorine, " "		38.6	38.1
Free ammonia " "		17.111	0.050
Albuminoid ammonia, " "		4.389	0.079
Nitrogen as nitrates, " "		0.110	9.220
Nitrogen as nitrites, " "		0.011	0.005
Bacteria, number per cubic centimeter,		633 000	120

Chemical precipitation and filter beds have been combined in order to reduce the area required for the beds or to increase the rate of filtration. As chemical precipitation destroys about 95 per cent of the bacteria, it seems desirable in such a case that the effluent from the precipitation tanks should be stored in basins for a sufficient length of time to enable the bacteria to increase to such numbers that they may do effective work in continuing the purification in the filter beds. Although this combination of methods has been tried in a number of towns, general deductions are lacking as to its economy and comparative efficiency. The works at East Orange, N. J., which were of this kind, were abandoned after a few years of use on account of the expense involved. Straining the sewage through coke before applying it to the filter beds has been used at Reading, Pa., and aeration is also secured by a double system of beds, the upper one being built on a steel superstructure. Of the pure system of intermittent filtration one of the best examples in America is that at Brockton, Mass.; the system is also in operation at Spencer, Mass., Pawtucket, R. I., Altoona, Pa., and several other places

72. Broad Irrigation.

Irrigation by water has been practiced from the earliest times in countries where the annual rainfall is less than 20 inches, the rainfall of the wet months being stored in reservoirs, from which it is distributed to the fields in the dry months by means of canals and ditches. The disposal of the sewage of towns by means of irrigation is, on the other hand, a modern method which was originated in England about 1870, and in which the principles of common irrigation are combined with those of intermittent filtration. It should be said, however, that the disposal of liquid kitchen wastes by running them into gardens has always been practiced by farmers, and indeed the universal custom of spreading the contents of privies and cesspools upon the fields is an imperfect application of the method on a small scale.

When a system of beds is prepared for the disposal of sewage by intermittent filtration vegetables may be planted upon some of the beds in the spring and only a small quantity of sewage be applied to them during the summer, the main work of purification being confined to the other beds. The beds which are planted are not those which have been especially prepared in layers of gravel and sand, but are areas on which the natural soil has been left undisturbed except in the places where trenches have been dug for the underdrains. The sewage is brought along the side of one of these beds in a ditch, from which it flows in lateral ditches or furrows between the rows of growing vegetables and is absorbed by the soil. The quantity of sewage thus furnished must be so regulated that the ground may not become wet enough to cause the growth of the vegetables to be unhealthy; this quantity can be determined only by experience, as it depends upon the kind of soil and growing plants and upon the amount of rainfall. The effect of the application of sewage is like that of manure; the organic matter is decomposed in the soil under the action of bacteria and the resulting products are absorbed by the roots of plants as materials for their growth.

Broad irrigation, or sewage farming, as it is often called, is the application of sewage upon fields not only for the purpose of disposing of it, but to utilize it as manure for the growing crops. The area needed to thus dispose of the sewage of a town is much larger than that required in the method of intermittent filtration, in fact from 10 to 20 times as large, but the expense of underdraining and embankments is much smaller per acre, and it is expected that a good profit will be derived from the sale of the crops.

The best location for a sewage farm is where the surface has a gentle slope toward a stream and where the soil is dry and porous. A part of the area should be laid out in specially prepared beds to be operated by the method of intermittent filtration when the sewage is not needed on the fields, but over the greater part the soil is left undisturbed except where underdrains are needed. The number of these drains will depend upon the character of the soil and the slope of the surface; moist and level ground must be well underdrained, but dry and sandy soil with a sloping surface needs few if any drains. Sometimes open ditches may serve as drains, but more commonly tiles are laid about five feet deep, the direction of the lines of tiles running normal to the contour curves of the surface so that their grades or slopes may be as great as possible.

The sewage is carried to the fields, if possible, by gravity, the main sewer from the town being extended nearly to the farm, where it changes into an open masonry conduit, in which a screen is placed to intercept the rags and coarse material. From this conduit the sewage passes through gates into ditches which lead to different parts of the farm, and from these ditches lateral furrows convey it between the rows of vegetables. All of these channels are laid out with respect to the contours of the surface, so that the flow may be neither too fast nor too slow, and many gates are provided for shutting off and regulating the discharge. These details, in fact, are practically the same as those which have been so long used in irrigating fields by water.

The application of sewage to the different fields must be intermittent, and often several days may elapse between the intervals of watering. While one part of the farm is watered another part is being planted, cultivated, or harvested, or is at rest. During periods of rainfall, or in the winter, when the fields need no sewage, it is turned upon the beds of the intermittent-filtration area. To secure success in the purification of the sewage and at the same time produce good crops constant and intelligent supervision of the processes of watering and resting is indispensable. The crops which are raised on sewage farms are not limited to vegetables, but wheat, oats, and grasses have been grown with success; in some cases three or four crops of grass have been obtained in one year where only one could be produced on unirrigated soil.

The operation of filtration areas in winter has not been found so difficult as might be expected. As the temperature of the sewage is higher than that of the air and soil, the ground is generally prevented from freezing, and hence the process of purification goes on, although the rate of filtration must be lower than in the summer, since the activity of the bacteria is not as great. In some cases special ditches have been made to receive the sewage in winter instead of spreading it over the surface, its absorption into the beds being through the bottom and sides of these ditches. In a severe climate where the temperature may be below 20 degrees Fahrenheit for several weeks more or less trouble will be experienced from freezing.

The largest example of broad irrigation is at Berlin, Germany, where an area of about 10 000 acres absorbs and utilizes the sewage of about 1 800 000 people. Throughout Germany, France, and England the system is extensively used and has been found to be a satisfactory one; the crops yield a fair return, the health of the farmers continues good, and the sewage of the cities is effectively turned into harmless and useful constituents. In America the system was first tried at Pullman, Ill., in 1887, but the level ground and dense soil was not such as to insure success and it was abandoned after a few years. At South Framingham, Mass., Meriden, Conn., Plainfield, N. J., Los Angeles, Cal., and at a few other places, such farms are in successful operation, and the system is regarded by the highest sanitary authorities as the one which is likely to secure the most effective disposal of sewage with the least expense.

73. CONCLUSION.

The various systems for the disposal of the organic wastes of a town have now been briefly described. It has been shown how fire, water, air, and earth, which have always been known to be effective destroyers of decaying organic matter in small quantities, may be applied in a scientific and economical manner to the disposal of the refuse and sewage of a large town. The manner is scientific because the reasons are known. Fire changes the organic matter into carbonaceous gases and mineral substances, water and air furnish oxygen by which the bacteria are enabled to perform a similar decomposition, and water and air acting in the earth do their work under conditions which lead to no offense. The processes which man has found best are after all the same processes that nature uses. Dirty water thrown upon sandy ground percolates into the earth and ultimately a part of it becomes pure ground water; this is the system of intermittent filtration. Foul water thrown over a cultivated field percolates and is absorbed by plants, and thus a part of it becomes ground water and another part becomes living organic matter; this is the system of broad irrigation. In both these systems a certain time is necessary for the bacteria to carry on their operations; if man attempts to shorten this time the results are unsatisfactory. By analyzing the effluent and properly regulating the rate of filtration he controls the process, so that complete purification is obtained.

The water-works of the town have long been in operation, bringing to all the inhabitants an abundant supply of clear, pure, and sparkling water. Now, after many years of agitation, the sewers have been laid, and from every house the sewage flows away to the fields where it is purified. Man has turned the water supply into sewage, now he turns the sewage back into pure water and growing vegetation The streets are clean, no filth accumulates around the dwellings, the air is pure, and the inhabitants are strong in body and mind. The second part of the construction work of the sanitary engineer is done.

Of all the branches of engineering that of sanitary engineering is the most interesting and important. It is interesting and fascinating by reason of its wide scope, wherein the results of the labors of the physician, the chemist, the biologist, the hydraulician, and the constructing engineer find application. It is important because it is work for the welfare of the community, and has its influence upon all surrounding communities and upon the nation. The city engineer who has built a water-works and a sewerage system finds in his completed work satisfaction and pride of a high degree, for through these the public health is promoted, the spread of disease is prevented, and therefore the world is rendered stronger and better.

In the last scene of the last act of the second part of Goethe's Faust the hero of that drama is represented as a sanitary engineer. All the dissipations and glories of the world had failed to extort from him an expression of satisfaction. In his old age, blind and feeble, he conceives the idea of building a wall near a village to guard the people from the storms of the sea, and of draining its soil to protect them from malarial disease. Mephistopheles acts as superintendent of the work and Faust hears the hum of its busy progress. He imagines the constructions complete and sees in the result a strong and free people. So glorious is the vision that he exclaims, " Verweile doch, du bist so schön!" and this utterance is the cause both of his bodily death and of his spiritual salvation.

Lo! the gates of the nineteenth century are about to close and those of the twentieth century to open. The mind of the sanitary engineer who has just completed his system of sewage removal and disposal turns back, for a moment only, to the dark ages preceding the seventeenth century, for the picture is one of horror. The slow awakening during the eighteenth century, when men were working to learn the truth, gives an image of hope and gladness. The gradual advance of the first part of the nineteenth century, and especially the wonderful progress since 1850, in which hygiene and sanitary science have played a part so great as to increase the median age of the American people by nearly four years, yields a picture of joyous brightness that causes his mind to fill with thanksgiving. Then arises a fair but indistinct vision of the advance in science, of the perfection in theory and practice, of the increase in health and longevity, and of that more glorious growth in mental and moral strength which the twentieth century shall witness.

74. Exercises and Problems.

62. Consult reports of the street departments of New York, Philadelphia, and Chicago, and ascertain the cost per mile of cleaning streets in different years; also the cost per inhabitant per year.

63 (*a*) From the same reports ascertain the average cost of removing garbage per person per year.

63 (*b*) How do the bacteria of putrefaction differ from the bacteria of ordinary decay?

64. Consult the technical journals since 1890, and obtain sketches and descriptions of the Rider and Engle furnaces for garbage cremation.

65 (*a*) Collect sketches and descriptions of the apparatus used in the Merz and Arnold systems of garbage digestion.

65 (*b*) Consult Engineering Record of August 6, 1898, and give accounts of the methods of garbage disposal in Glasgow, Scotland, and Buffalo, N. Y.

66. Make an estimate of the cost of construction and operation of a cesspool 20 feet deep to hold the sewage of 6 people.

67 (*a*) Consult Clark's article on sewage in Engineering News of August 4, 1898, and explain the septic tank method for the purification of sewage.

67 (*b*) Read Hering's article on Bacterial Processes of Sewage Purification in The Engineering Magazine for September, 1898, and explain the modern theory of septic decomposition.

68 (*a*) Consult Sedgwick's article in Report of State Board of Health of Massachusetts for 1892, and give details regarding the transmission of typhoid fever germs in the Merrimac River from Lowell to Newburyport.

68 (*b*) Explain how the great drainage canal at Chicago is to purify the sewage of that city.

69 (*a*) Describe the method of purifying sewage by screening through hay and sand which was used at Atlantic City, N. J., in 1893.

69 (*b*) Describe the combined screening, aeration, and filtration plant constructed at Reading, Pa., in 1897.

70 (*a*) If 100 pounds of lime are added to water containing sufficient carbon dioxide to completely react with it, how many pounds of calcium carbonate will be precipitated?

70 (*b*) Consult Rafter and Baker's Sewage Disposal in the United States (New York, 1895), and describe the methods of chemical precipitation at Chautauqua, N. Y., and Canton, O.

71. Describe the construction work of the intermittent-filtration beds at Brockton, Mass.

72 (*a*) Consult Rafter's Sewage Irrigation (U. S. Geological Survey, Washington, 1897), and describe the sewage farms at Berlin and Paris.

72 (*b*) Describe the system of broad irrigation as practiced at Bristol, Conn., and Plainfield, N. J.

73. Consult Engineering Record, March 11, 1899, and describe the sewage disposal system of the Southern Indiana Hospital.

INDEX.

Aeration of filter beds, 82, 207
    sewage, 197, 200
    water, 63, 112
Air, composition of, 26
    in sewers, 166
Air inlet, 157
    lift pump, 116
Albumenoid ammonia, 32, 35
Algae, 37
Alum in precipitation, 70, 201
Aluminum, 72
Ammonia, 22, 32, 35, 185, 208
Analyses of sewage, 41, 193, 208
    water, 41, 65, 68, 76, 80
Analysis, biological, 36, 39
    chemical, 33, 39
Anderson's purifier, 77
Annual rainfall, 50
Aqueducts, 10, 103, 104, 106
Arched culverts, 103
Archimedian screw, 170
Artesian wells, 59, 60
Artificial filtration, 68, 70, 77, 205
Ashes, disposal of, 183
Assessments for sewers, 175

Babylon, reservoirs of, 9
Bacteria, 18, 23, 38, 40, 197
    classification, 19
    in sewage, 193, 199, 207
    in water, 38, 41, 80
Bacteriology, 18
Bacteriological analyses, 36, 37, 40
Basins for filtration, 67
    precipitation, 202

Ball trap, 143, 172
Bell trap, 143
Berlier's evacuator, 172
Berlin, sewage farms, 212
Bethlehem, Pa., water works, 87
Black death, 10
Boards of health, 15, 25, 197
Boiler for kitchen, 133
Boiling of water, 69
Boston, Mass., sewage disposal, 169
    water works, 87
Broad irrigation, 209
Bromine as disinfectant, 113
Brook water, 55
Brooklyn, N. Y., water supply 58

Canals, 103
Capacity of reservoirs, 91, 93
Cast-iron pipes, 107, 129
Catch basins, 155
Centrifugal pumps, 169
Cesspools, 53, 141, 190
Charcoal, 72, 198
Charleston, S. C., artesian wells, 59
Chemical analysis, 33, 36
    precipitation, 70, 200
Chloride of lime, 191
Chlorine, 33, 39, 57, 67, 208
Cholera, 24, 28, 40
    at Hamburg, 29
    at Manilla, 24
Chicago, sewerage, 169
    water supply, 56, 125
Cisterns, 53, 54
Civil engineering, 8

218　INDEX.

Classification of bacteria, 19
　　　　　diseases, 12
　　　　　sewerage systems, 149
　　　　　water works, 86
Cleaning of cesspools, 190
　　　　　cisterns, 54
　　　　　filter beds, 81, 207
　　　　　sewers, 165
　　　　　streets, 180
Clark's softening process, 69
Cloaca maxima, 140, 162
Coal, 117, 119
Coefficients for aqueducts, 105
　　　　　pipes, 109
　　　　　sewers, 169
Coke, 73, 169, 198
Collecting reservoirs, 60, 61
Collection of water samples, 33
Combined system of sewerage, 151, 167, 175
Concrete dams, 100
　　　　　foundations, 163
　　　　　sewers, 164
Conduits, 104, 202
Consumption, 13, 17
Consumption of water, 10, 89
Contagious diseases, 14
Continuous filtration, 82, 198
Copperas, 201
Core of a dam, 94
Covered filter beds, 82
　　　　　reservoirs, 62
Cremation of garbage, 186
Croton aqueducts, 106
Culture for bacteria, 38
Culverts, 103

D trap, 143
Damascus, water supply, 140
Dams, earthen, 94
　　　　　masonry, 97
Danzig, typhoid fever at, 24
Deaths in 1890, in U. S., 17
　　　　　black, 10

Deaths, registration of, 15
Decay of garbage, 185
　　　　　organic matter, 22
　　　　　sewage, 193
Deep wells, 59
Deodorizers, 191
Deposits in reservoirs, 112
　　　　　sewers, 154, 161
Decomposition of sewage, 192
Desmids, 37
Diatoms, 37
Digestion of garbage, 188
Diphtheria, 17, 192
Direct pumping, 88, 122, 125
Disinfectants, 191
Disk meter, 133
Disposal of garbage, 180, 189
　　　　　sewage, 190, 215
　　　　　sludge, 204
Distillation of water, 68
Disease and air, 25
　　　　　filth, 23
　　　　　water, 28
Diseases, classification of, 12
　　　　　germs of, 20
　　　　　zymotic, 14, 17,
Dissolved matter in water, 31
Distributing reservoirs, 62, 110, 113
Double acting pump, 114
Drainage of houses, 146
　　　　　soil, 152, 163
Drain pipes, 148, 210
Driven wells, 58
Duplex pump, 114
Duty of pumping engine, 117

Earth closet, 141
　　　　　dams, 94, 95
East Jersey, N. J., conduit, 107
　　　　　Orange, N. J., filter beds, 209
Effluent of filtration, 80, 202, 205
Electrical purification, 70, 85
Endemic diseases, 14
Epidemic diseases, 14
Epidemics, cholera, 22, 29

… INDEX. 219

Epidemics, black death, 10, 24
    typhoid fever, 24, 28, 29, 84, 190
Evaporation, 51
Exercises, 44, 84, 135, 178, 215

Factories, wastes of, 194
Failure of dams, 96
    stand pipes, 128
Farms, sewage, 210
Filter basin, 67
    beds, 78, 80, 83, 113 206
    galleries, 67, 85
Filters, household, 72
    mechanical, 70, 75
    Pasteur, 73
Filtration, artificial, 77, 83
    continuous, 82
    intermittent, 82, 205
    natural, 66
    slow sand, 77, 85
Filth and disease, 23
Fire hose, 130
    pressure, 130
    service, 90
    streams, 131
Flies, disease spread by, 190
Flush tanks, 134, 158
Foundations, 97, 162
Free ammonia, 32, 35
Friction in pipes, 109, 121
Frost protection for hydrants, 130
    dams, 95
Furnace for garbage, 186

Galleries for filtration, 67, 85
Garbage, collection of, 141, 183
    cremation of, 186
    definition of, 139
    decay of, 185
    digestion of, 188
Gases in sewage, 166, 193
    water, 31
Gate chambers, 103, 112
Germs of disease, 13, 19

Gravity sewerage systems, 151
    water supply systems, 86, 92, 113
Grease trap, 144
Ground water, 52, 57, 58

Hamburg, cholera at, 29
    filter beds of, 85
Hand hole in pipes, 157
Hard water, 32, 69
Hardness, scale of, 34
Health, 6, 12
Heisch's test of water, 38
Historical notes, 9, 139
House drainage, 146, 147
    filters, 72
    fixtures, 143
    meters, 133
    pipes, 132
    wells, 59
Hydraulics, 101, 105, 108, 121, 160
Hydraulic grade line, 108
    radius, 105, 108
Impure air, 25
    water, 28
India, sanitary condition, 25
Infectious diseases, 14
Influenza, 14
Intermittent filtration, 82, 199, 206
Interpretation of analyses, 39
Irrigation, 60, 209
Iron, precipitation by, 71, 77, 201
    perchloride, 71
    spongy, 73

Jerusalem, aqueducts of, 9
Johnstown, Pa., failure of dam, 96

Kitchen boiler, 134
    sink, 134, 144
Karcha, cisterns of, 9

Lake water, 56
Lamp hole, 157
Laundry fixtures, 134, 144

INDEX.

Lawrence, Mass., filter beds, 82
        typhoid fever, 79
Lead pipes, 133
Liernur's sewerage system, 171
Lime, 69, 71, 184, 201
Limewater, 69
London, drains, 11
    covered reservoir, 62
    mortality of, 18, 24
    sewerage, 141

Malaria, 13, 14, 26
Manholes, 155, 157, 166
Manila, cholera at, 24
Masonry dams, 98
    for sewers, 163
Mechanical filters, 70, 74
Median age of population, 43
Memphis, Tenn., sewers, 156, 178
Missouri river water, 63
Mortality, rate of, 15, 16

Natural filtration, 66
New York, aqueducts, 106
    water consumption, 93
    storage capacity, 110
Nitrates, 22, 33, 36, 185
Nitrification, 23, 36
Nitrites, 33, 185
Nitrogen as nitrates and nitrites, 36

Organic matter, 21, 23
Overflow sewers, 154
Oxygen, 11, 20, 26, 31
Oxygen consumed, 36

Pail system of removal, 141
Pasteur filter, 73
Pavements, 180
Penstocks, 167
Percolation, 51
Permanent hardness, 32
Paris, covered reservoirs, 62
    sewers, 142
Plants, growth of, 22

Pipes, 106, 107, 129, 133
    friction in, 121
    house, 132
    thickness of, 121, 127
Piston meter, 133
Pneumatic cart, 191
Pollution of reservoirs, 113
    rivers, 56, 195
Potassium permanganate, 71
Precipitation of sewage, 201, 202
    water, 70, 77
Pressure in pipes, 91, 131
    regulator, 131
Privies, 141, 190
Problems, 44, 84, 135, 178, 215
Pullman, Ill., sewage farm, 212
Purification of sewage, 192-213
    water, 47-82
Purity, standards of, 40
Putrefaction, 185
Pumps, 113
Pumping, direct, 122
    to reservoir, 120
    to stand-pipe, 124
    to tank, 123
Pumping engines, 116
    of sewage, 151, 168, 170

Rainfall, 47, 148
Rain gage, 47, 84
    water, 52, 53
Rate of filtration, 78, 206
    mortality, 16
Reading, Pa., water supply, 64
    coke filtration, 209
Regulation of pressure, 131
Removal of garbage, 63, 183
    refuse, 182
Reservoir dams, 94, 97, 100
    embankments, 94, 112
Reservoirs, ancient, 60
    distributing, 62, 86, 110
    storage, 61, 86
    pumping, 2, 120
River pollution, 56, 195

INDEX.

River water, 55
Rivers, purification of, 195
Rochester reservoir, 62
    water-pipe, 107
Rome, sewers, 140
    aqueducts, 10
Rotary meter, 133
    sweeper, 183
Runoff, 51

S trap, 143
Sanitary code, 9, 44
    engineering, 9, 12, 214
    science, 7–42
Sand for filter beds, 81
Schuylkill river, 196
Scraping streets, 181
Screens, 72, 197, 169
Sedimentation, 63, 84, 199
Separate system of sewerage, 151, 156, 167
Septic decomposition, 183
    sewage, 193, 207
    tank method, 215
Sewage, 139, 195–213
    aeration, 197
    analyses, 41, 193, 150
    filtration, 198, 210
    precipitation, 200, 203
    pumping, 168, 171
    screening, 169, 197
    storage, 199
Sewerage systems, 139–178
    classification, 149
    combined, 152
    separate, 156
    vacuum, 170
Sewer gas, 155, 166
Sewers, ancient, 140
    clearing of, 165
    construction of, 162
    sizes of, 159
    ventilation of, 162
Shone system, 173
Silt, 31, 63

Single-acting pump, 162
Slopes of sewers, 160
Sludge, 202, 204
Small pox, 11, 14
Snow, 54
Sodium chloride, 32
Softening of hard water, 69
Soil pipe, 146
Spongy iron, 73
Spring water, 57
Springs, 57, 60, 190
Stale sewage, 193
Standards of purity, 40
Standpipes, 124, 126, 127
Steam engines, 117
Storage of sewage, 199
    water, 63
Storage reservoirs, 60, 91, 103, 107
Street cleaning, 180
    dust, 26
    pavements, 180
    mains, 129
Strainers, 72
Suction pump, 113
Sunlight, action on bacteria, 20
Surface water, 54
Suspended matter in water, 30
Swamp water, 55
Sweeping streets, 181
Syphon flush tank, 159
Syphonage of traps, 144

Tables, list of :
    Aqueducts, 105
    Deaths in U. S., 17
    Median ages, 43
    Pipes, 109, 121
    Pumping engines, 118
    Rainfall, 50
    Sewers, 160
Tank reservoirs, 123, 126
Tankage, 188
Tanks for precipitation, 203
Temporary hardness, 32
Thickness of dams, 98

Thickness of pipes, 121, 127
Tilting flush tank, 159
Topography, 92, 153
Total solids in sewage, 150, 192
        water, 34
Traps, 143, 146, 166
Triplex pump, 114
Tuberculosis, 13, 26
Typhoid fever, cause, 13, 20. 28
    epidemics, 13, 28, 190
    deaths from, 17, 79
Typical analyses, 41

Vaccination, 11, 14
Vacuum pump, 113
Vacuum systems of sewerage, 151, 170
Vaults for filter beds, 82
Velocity in pipes, 109
    in sewers, 161
    of sedimentation, 63
Vent tube, 147
Ventilation of houses, 27

Ventilation of sewers, 166
    streets, 27
Venturi water meter, 132

Waste weirs, 97, 101
Water, analyses, 41, 65, 68, 76, 80
    biological analysis, 37, 39
    chemical analysis, 33, 39
    closets, 134, 145
    consumption of, 89
    matter in, 30
    meters, 132
    purification, 47-85
Watershed, 51, 92. 102
Water-supply systems, 86-138
Water works in U. S., 88
Wells, 58, 59, 88, 190
Wetted perimeter, 105
Williamsburg, Mass., failure of dam, 96
World's Fair, sewerage of, 174

Zymotic diseases, 13, 17, 24, 28, 37

WORKS of Professor MANSFIELD MERRIMAN

PUBLISHED BY

JOHN WILEY & SONS,

53 East Tenth Street, New York.

LONDON: CHAPMAN & HALL, LIMITED.

Elements of Sanitary Engineering. For the use of Engineering Students and Municipal Officers. Octavo, cloth. Price $2.00.

Treatise on Hydraulics. A Text-book for Students and a Manual for Engineers. Octavo, cloth. Price $4.00.

Mechanics of Materials and of Beams, Columns, and Shafts. Octavo, cloth. Enlarged edition. Price $4.00.

Text-book on the Method of Least Squares. Octavo, cloth. Enlarged edition. Price $2.00.

Strength of Materials. An Elementary Text-book for Manual-training Schools. Duodecimo, cloth. Price $1.00.

Text-book on Roofs and Bridges. By Professors MERRIMAN and JACOBY. In Four Parts. Octavo, cloth.
- PART I. **Stresses in Simple Trusses.** Price $2.50.
- PART II. **Graphic Statics.** Price $2.50.
- PART III. **Bridge Design.** Price $2.50.
- PART IV. **Higher Structures.** Price $2.50.

Higher Mathematics. A Text-book for Classical and Engineering Colleges. Edited by Professors MERRIMAN and WOODWARD. Octavo, cloth. Price $5.00.

SHORT-TITLE CATALOGUE

OF THE

PUBLICATIONS

OF

JOHN WILEY & SONS,

NEW YORK.

LONDON: CHAPMAN & HALL, LIMITED.

ARRANGED UNDER SUBJECTS.

Descriptive circulars sent on application.
Books marked with an asterisk are sold at *net* prices only.
All books are bound in cloth unless otherwise stated.

AGRICULTURE.

CATTLE FEEDING—DAIRY PRACTICE—DISEASES OF ANIMALS—GARDENING, ETC.

Armsby's Manual of Cattle Feeding............12mo,	$1	75
Downing's Fruit and Fruit Trees........................8vo,	5	00
Grotenfelt's The Principles of Modern Dairy Practice. (Woll.) 12mo,	2	00
Kemp's Landscape Gardening.... ....12mo,	2	50
Loudon's Gardening for Ladies. (Downing.)............12mo,	1	50
Maynard's Landscape Gardening..................12mo,	1	50
Steel's Treatise on the Diseases of the Dog..............8vo,	3	50
" Treatise on the Diseases of the Ox..................8vo,	6	00
Stockbridge's Rocks and Soils.... ....8vo,	2	50
Woll's Handbook for Farmers and Dairymen............12mo,	1	50

ARCHITECTURE.

BUILDING—CARPENTRY—STAIRS—VENTILATION—LAW, ETC.

Berg's Buildings and Structures of American Railroads.....4to,	7	50
Birkmire's American Theatres—Planning and Construction.8vo,	3	00
" Architectural Iron and Steel..................8vo,	3	50
" Compound Riveted Girders..................8vo,	2	00
" Skeleton Construction in Buildings 8vo,	3	00

Birkmire's Planning and Construction of High Office Buildings.
8vo, $3 50
Carpenter's Heating and Ventilating of Buildings..........8vo, 3 00
Freitag's Architectural Engineering.................8vo, 2 50
Gerhard's Sanitary House Inspection....................16mo, 1 00
" Theatre Fires and Panics......................12mo, 1 50
Hatfield's American House Carpenter.....................8vo, 5 00
Holly's Carpenter and Joiner.. ....18mo, 75
Kidder's Architect and Builder's Pocket-book...16mo, morocco, 4 00
Merrill's Stones for Building and Decoration..............8vo, 5 00
Monckton's Stair Building—Wood, Iron, and Stone........4to, 4 00
Wait's Engineering and Architectural Jurisprudence......8vo, 6 00
Sheep, 6 50
Worcester's Small Hospitals—Establishment and Maintenance, including Atkinson's Suggestions for Hospital Architecture... ....12mo, 1 25
World's Columbian Exposition of 1893.............Large 4to, 2 50

ARMY, NAVY, Etc.

MILITARY ENGINEERING—ORDNANCE—LAW, ETC.

Bourne's Screw Propellers...............................4to, 5 00
*Bruff's Ordnance and Gunnery........................8vo, 6 00
Chase's Screw Propellers................................8vo, 3 00
Cooke's Naval Ordnance 8vo, 12 50
Cronkhite's Gunnery for Non-com. Officers.....32mo, morocco, 2 00
*Davis's Treatise on Military Law......................8vo, 7 00
Sheep, 7 50
* " Elements of Law...............................8vo, 2 50
De Brack's Cavalry Outpost Duties. (Carr.)....32mo, morocco, 2 00
Dietz's Soldier's First Aid.....................16mo, morocco, 1 25
*Dredge's Modern French Artillery....Large 4to, half morocco, 15 00
" Record of the Transportation Exhibits Building, World's Columbian Exposition of 1893..4to, half morocco, 10 00
Durand's Resistance and Propulsion of Ships.............8vo, 5 00
Dyer's Light Artillery....................................12mo, 3 00
Hoff's Naval Tactics......................................8vo, 1 50
*Ingalls's Ballistic Tables................................8vo, 1 50
" Handbook of Problems in Direct Fire............8vo, 4 00

Mahan's Permanent Fortifications. (Mercur.).8vo, half morocco,	$7 50
Mercur's Attack of Fortified Places....................12mo,	2 00
" Elements of the Art of War.....................8vo,	4 00
Metcalfe's Ordnance and Gunnery..........12mo, with Atlas,	5 00
Murray's A Manual for Courts-Martial........16mo, morocco,	1 50
" Infantry Drill Regulations adapted to the Springfield Rifle, Caliber .45.....................32mo, paper,	10
*⁢Phelps's Practical Marine Surveying....................8vo,	2 50
Powell's Army Officer's Examiner......................12mo,	4 00
Sharpe's Subsisting Armies..................32mo, morocco,	1 50
Very's Navies of the World...............8vo, half morocco,	3 50
Wheeler's Siege Operations..............................8vo,	2 00
Winthrop's Abridgment of Military Law..............12mo,	2 50
Woodhull's Notes on Military Hygiene.................16mo,	1 50
Young's Simple Elements of Navigation.......16mo, morocco,	2 00
" " " " " first edition........	1 00

ASSAYING.

Smelting—Ore Dressing—Alloys, Etc.

Fletcher's Quant. Assaying with the Blowpipe..16mo, morocco,	1 50
Furman's Practical Assaying............................8vo,	3 00
Kunhardt's Ore Dressing...............................8vo,	1 50
O'Driscoll's Treatment of Gold Ores.....................8vo,	2 00
Ricketts and Miller's Notes on Assaying.................8vo,	3 00
Thurston's Alloys, Brasses, and Bronzes.............. ..8vo,	2 50
Wilson's Cyanide Processes...........................12mo,	1 50
" The Chlorination Process....................12mo,	1 50

ASTRONOMY.

Practical, Theoretical, and Descriptive.

Craig's Azimuth..4to,	3 50
Doolittle's Practical Astronomy.........................8vo,	4 00
Gore's Elements of Geodesy............................8vo,	2 50
Hayford's Text-book of Geodetic Astronomy............8vo.	3 00
* Michie and Harlow's Practical Astronomy..............8vo,	3 00
* White's Theoretical and Descriptive Astronomy........12mo,	2 00

BOTANY.

GARDENING FOR LADIES, ETC.

Baldwin's Orchids of New England...............Small 8vo,	$1 50
Loudon's Gardening for Ladies. (Downing.)...........12mo,	1 50
Thomé's Structural Botany......................16mo,	2 25
Westermaier's General Botany. (Schneider.).............8vo,	2 00

BRIDGES, ROOFS, Etc.

CANTILEVER—DRAW—HIGHWAY—SUSPENSION.

(See also ENGINEERING, p. 7.)

Boller's Highway Bridges....................................8vo,	2 00
* " The Thames River Bridge..................4to, paper,	5 00
Burr's Stresses in Bridges......................................8vo,	3 50
Crehore's Mechanics of the Girder........................8vo,	5 00
Dredge's Thames Bridges................7 parts, per part,	1 25
Du Bois's Stresses in Framed Structures.............Small 4to,	10 00
Foster's Wooden Trestle Bridges..........................4to,	5 00
Greene's Arches in Wood, etc.............................8vo,	2 50
" Bridge Trusses....................................8vo,	2 50
" Roof Trusses.....................................8vo,	1 25
Howe's Treatise on Arches 8vo,	4 00
Johnson's Modern Framed Structures..............Small 4to,	10 00
Merriman & Jacoby's Text-book of Roofs and Bridges. Part I., Stresses..8vo,	2 50
Merriman & Jacoby's Text-book of Roofs and Bridges. Part II., Graphic Statics 8vo,	2 50
Merriman & Jacoby's Text-book of Roofs and Bridges. Part III., Bridge Design...........................8vo,	2 50
Merriman & Jacoby's Text-book of Roofs and Bridges. Part IV., Continuous, Draw, Cantilever, Suspension, and Arched Bridges.....................................8vo,	2 50
* Morison's The Memphis Bridge..................Oblong 4to,	10 00
Waddell's Iron Highway Bridges.........................8vo,	4 00
" De Pontibus (a Pocket-book for Bridge Engineers). 16mo, morocco,	3 00
Wood's Construction of Bridges and Roofs................8vo,	2 00
Wright's Designing of Draw Spans. Parts I. and II..8vo, each	2 50
" " " " " Complete...........8vo,	3 50

CHEMISTRY.

QUALITATIVE—QUANTITATIVE—ORGANIC—INORGANIC, ETC.

Adriance's Laboratory Calculations..................12mo,	$1 25
Allen's Tables for Iron Analysis......................8vo,	3 00
Austen's Notes for Chemical Students.................12mo,	1 50
Bolton's Student's Guide in Quantitative Analysis.........8vo,	1 50
Classen's Analysis by Electrolysis. (Herrick and Boltwood.).8vo,	3 00
Crafts's Qualitative Analysis. (Schaeffer.)..............12mo,	1 50
Drechsel's Chemical Reactions. (Merrill.)..............12mo,	1 25
Fresenius's Quantitative Chemical Analysis. (Allen.).......8vo,	6 00
" Qualitative " " (Johnson.).....8vo,	3 00
" " " " (Wells.) Trans. 16th German Edition..............................8vo,	5 00
Fuertes's Water and Public Health....................12mo,	1 50
Gill's Gas and Fuel Analysis........................12mo,	1 25
Hammarsten's Physiological Chemistry. (Mandel.).........8vo,	4 00
Helm's Principles of Mathematical Chemistry. (Morgan).12mo,	1 50
Kolbe's Inorganic Chemistry.........................12mo,	1 50
Ladd's Quantitative Chemical Analysis.................12mo,	1 00
Landauer's Spectrum Analysis. (Tingle.)................8vo,	3 00
Löb's Electrolysis and Electrosynthesis of Organic Compounds. (Lorenz.)..12mo,	1 00
Mandel's Bio-chemical Laboratory....................12mo,	1 50
Mason's Water-supply..............................8vo,	5 00
" Examination of Water.......................12mo,	1 25
Meyer's Organic Analysis. (Tingle.) (*In the press.*)	
Miller's Chemical Physics...........................8vo,	2 00
Mixter's Elementary Text-book of Chemistry............12mo,	1 50
Morgan's The Theory of Solutions and its Results........12mo,	1 00
" Elements of Physical Chemistry...............12mo,	2 00
Nichols's Water-supply (Chemical and Sanitary)...........8vo,	2 50
O'Brine's Laboratory Guide to Chemical Analysis..........8vo,	2 00
Perkins's Qualitative Analysis.......................12mo,	1 00
Pinner's Organic Chemistry. (Austen.)................12mo,	1 50
Poole's Calorific Power of Fuels.......................8vo,	3 00
Ricketts and Russell's Notes on Inorganic Chemistry (Non-metallic)................................Oblong 8vo, morocco,	75
Ruddiman's Incompatibilities in Prescriptions............8vo,	2 00

Schimpf's Volumetric Analysis.........................12mo,	$2	50
Spencer's Sugar Manufacturer's Handbook.....16mo, morocco,	2	00
" Handbook for Chemists of Beet Sugar Houses. 16mo, morocco,	3	00
Stockbridge's Rocks and Soils............................8vo,	2	50
Tillman's Descriptive General Chemistry. (*In the press.*)		
Van Deventer's Physical Chemistry for Beginners. (Boltwood.) 12mo,	1	50
Wells's Inorganic Qualitative Analysis..................12mo,	1	50
" Laboratory Guide in Qualitative Chemical Analysis. 8vo,	1	50
Whipple's Microscopy of Drinking-water.................8vo,	3	50
Wiechmann's Chemical Lecture Notes.....................12mo,	3	00
" Sugar Analysis....................Small 8vo,	2	50
Wulling's Inorganic Phar. and Med. Chemistry...........12mo,	2	00

DRAWING.

ELEMENTARY—GEOMETRICAL—MECHANICAL—TOPOGRAPHICAL.

Hill's Shades and Shadows and Perspective..............8vo,	2	00
MacCord's Descriptive Geometry.........................8vo,	3	00
" Kinematics...................................8vo,	5	00
" Mechanical Drawing..........................8vo,	4	00
Mahan's Industrial Drawing. (Thompson.)........2 vols., 8vo,	3	50
Reed's Topographical Drawing. (H. A.).................4to,	5	00
Reid's A Course in Mechanical Drawing..................8vo,	2	00
" Mechanical Drawing and Elementary Machine Design. 8vo. (*In the press.*)		
Smith's Topographical Drawing. (Macmillan.)...........8vo,	2	50
Warren's Descriptive Geometry....................2 vols., 8vo,	3	50
" Drafting Instruments.........................12mo,	1	25
" Free-hand Drawing...........................12mo,	1	00
" Linear Perspective...........................12mo,	1	00
" Machine Construction...................2 vols., 8vo,	7	50
" Plane Problems...............................12mo,	1	25
" Primary Geometry............................12mo,		75
" Problems and Theorems.......................8vo,	2	50
" Projection Drawing..........................12mo,	1	50

Warren's Shades and Shadows.........8vo,	$3	00
" Stereotomy—Stone-cutting.........8vo,	2	50
Whelpley's Letter Engraving.........12mo,	2	00

ELECTRICITY AND MAGNETISM.

Illumination—Batteries—Physics—Railways.

Anthony and Brackett's Text-book of Physics. (Magie.) Small 8vo,	3	00
Anthony's Theory of Electrical Measurements.........12mo,	1	00
Barker's Deep-sea Soundings.........8vo,	2	00
Benjamin's Voltaic Cell.........8vo,	3	00
" History of Electricity.........8vo,	3	00
Classen's Analysis by Electrolysis. (Herrick and Boltwood.) 8vo,	3	00
Cosmic Law of Thermal Repulsion.........12mo,		75
Crehore and Squier's Experiments with a New Polarizing Photo-Chronograph.........8vo,	3	00
Dawson's Electric Railways and Tramways. Small, 4to, half morocco,	12	50
* Dredge's Electric Illuminations....2 vols., 4to, half morocco,	25	00
" " " Vol. II.........4to,	7	50
Gilbert's De magnete. (Mottelay.).........8vo,	2	50
Holman's Precision of Measurements.........8vo,	2	00
" Telescope-mirror-scale Method.........Large 8vo,		.75
Löb's Electrolysis and Electrosynthesis of Organic Compounds. (Lorenz.).........12mo,	1	00
*Michie's Wave Motion Relating to Sound and Light.........8vo,	4	00
Morgan's The Theory of Solutions and its Results.........12mo,	1	00
Niaudet's Electric Batteries. (Fishback.).........12mo,	2	50
Pratt and Alden's Street-railway Road-beds.........8vo,	2	00
Reagan's Steam and Electric Locomotives.........12mo,	2	00
Thurston's Stationary Steam Engines for Electric Lighting Purposes.........8vo,	2	50
*Tillman's Heat.........8vo,	1	50

ENGINEERING.

Civil—Mechanical—Sanitary, Etc.

(*See also* Bridges, p. 4; Hydraulics, p. 9; Materials of Engineering, p. 10; Mechanics and Machinery, p. 12; Steam Engines and Boilers, p. 14.)

Baker's Masonry Construction..........................8vo,	$5 00
" Surveying Instruments......................12mo,	3 00
Black's U. S. Public Works....................Oblong 4to,	5 00
Brooks's Street-railway Location..............16mo, morocco,	1 50
Butts's Civil Engineers' Field Book...........16mo, morocco,	2 50
Byrne's Highway Construction..........................8vo,	5 00
" Inspection of Materials and Workmanship.......16mo,	3 00
Carpenter's Experimental Engineering 8vo,	6 00
Church's Mechanics of Engineering—Solids and Fluids....8vo,	6 00
" Notes and Examples in Mechanics..............8vo,	2 00
Crandall's Earthwork Tables...........................8vo,	1 50
" The Transition Curve..............16mo, morocco,	1 50
*Dredge's Penn. Railroad Construction, etc. Large 4to, half morocco,	20 00
* Drinker's Tunnelling...................4to, half morocco,	25 00
Eissler's Explosives—Nitroglycerine and Dynamite........8vo,	4 00
Folwell's Sewerage......................................8vo,	3 00
Fowler's Coffer-dam Process for Piers...................8vo.	2 50
Gerhard's Sanitary House Inspection....................12mo,	1 00
Godwin's Railroad Engineer's Field-book......16mo, morocco,	2 50
Gore's Elements of Geodesy.......... ....8vo,	2 50
Howard's Transition Curve Field-book.........16mo, morocco,	1 50
Howe's Retaining Walls (New Edition.)..................12mo,	1 25
Hudson's Excavation Tables. Vol. II..................8vo,	1 00
Hutton's Mechanical Engineering of Power Plants........8vo,	5 00
Johnson's Materials of Construction...............Large 8vo,	6 00
" Stadia Reduction Diagram..Sheet, $22\frac{1}{4} \times 28\frac{1}{4}$ inches,	50
" Theory and Practice of Surveying........Small 8vo,	4 00
Kent's Mechanical Engineer's Pocket-book.....16mo, morocco,	5 00
Kiersted's Sewage Disposal.............................12mo,	1 25
Mahan's Civil Engineering. (Wood.)....................8vo,	5 00
Merriman and Brook's Handbook for Surveyors....16mo, mor.,	2 00
Merriman's Geodetic Surveying..........................8vo,	2 00
" Retaining Walls and Masonry Dams............8vo,	2 00
" Sanitary Engineering..........................8vo,	2 00
Nagle's Manual for Railroad Engineers........16mo, morocco,	3 00
Ogden's Sewer Design. 12mo,	2 00
Patton's Civil Engineering................8vo, half morocco,	7 50

Patton's Foundations..8vo,	$5	00
Pratt and Alden's Street-railway Road-beds..............8vo,	2	00
Rockwell's Roads and Pavements in France........ ....12mo,	1	25
Searles's Field Engineering ....16mo, morocco,	3	00
" Railroad Spiral...................16mo, morocco,	1	50
Siebert and Biggin's Modern Stone Cutting and Masonry...8vo,	1	50
Smart's Engineering Laboratory Practice................12mo,	2	50
Smith's Wire Manufacture and Uses..............Small 4to,	3	00
Spalding's Roads and Pavements..... ....12mo,	2	00
" Hydraulic Cement...........................12mo,	2	00
Taylor's Prismoidal Formulas and Earthwork............8vo,	1	50
Thurston's Materials of Construction 8vo,	5	00
* Trautwine's Civil Engineer's Pocket-book....16mo, morocco,	5	00
* " Cross-section...........................Sheet,		25
* " Excavations and Embankments.............8vo,	2	00
* " Laying Out Curves.............12mo, morocco,	2	50
Waddell's De Pontibus (A Pocket-book for Bridge Engineers). 16mo, morocco,	3	00
Wait's Engineering and Architectural Jurisprudence.......8vo,	6	00
Sheep,	6	50
" Law of Field Operation in Engineering, etc........8vo.		
Warren's Stereotomy—Stone-cutting....................8vo,	2	50
*Webb's Engineering Instruments.............16mo, morocco,		50
" " " New Edition.............	1	25
Wegmann's Construction of Masonry Dams..............4to,	5	00
Wellington's Location of Railways...............Small 8vo,	5	00
Wheeler's Civil Engineering...........................8vo,	4	00
Wolff's Windmill as a Prime Mover....................8vo,	3	00

HYDRAULICS.

WATER-WHEELS—WINDMILLS—SERVICE PIPE—DRAINAGE, ETC.

(*See also* ENGINEERING, p. 7.)

Bazin's Experiments upon the Contraction of the Liquid Vein. (Trautwine.)..8vo,	2	00
Bovey's Treatise on Hydraulics........................8vo,	4	00
Coffin's Graphical Solution of Hydraulic Problems......12mo,	2	50
Ferrel's Treatise on the Winds, Cyclones, and Tornadoes...8vo,	4	00
Fuertes's Water and Public Health....................12mo,	1	50
Ganguillet & Kutter's Flow of Water. (Hering & Trautwine.) 8vo,	4	00
Hazen's Filtration of Public Water Supply...............8vo,	2	00
Herschel's 115 Experiments 8vo,	2	00

Kiersted's Sewage Disposal............................12mo,	$1	25
Mason's Water Supply..8vo,	5	00
" Examination of Water............................12mo,	1	25
Merriman's Treatise on Hydraulics........................8vo,	4	00
Nichols's Water Supply (Chemical and Sanitary)..........8vo,	2	50
Wegmann's Water Supply of the City of New York........4to,	10	00
Weisbach's Hydraulics. (Du Bois.).......................8vo,	5	00
Whipple's Microscopy of Drinking Water..................8vo,	3	50
Wilson's Irrigation Engineering..........................8vo,	4	00
" Hydraulic and Placer Mining....................12mo,	2	00
Wolff's Windmill as a Prime Mover.......................8vo,	3	00
Wood's Theory of Turbines................................8vo,	2	50

MANUFACTURES.

BOILERS—EXPLOSIVES—IRON—STEEL—SUGAR—WOOLLENS, ETC.

Allen's Tables for Iron Analysis..........................8vo,	3	00
Beaumont's Woollen and Worsted Manufacture.........12mo,	1	50
Bolland's Encyclopædia of Founding Terms............12mo,	3	00
" The Iron Founder..............................12mo,	2	50
" " " " Supplement.................12mo,	2	50
Bouvier's Handbook on Oil Painting....................12mo,	2	00
Eissler's Explosives, Nitroglycerine and Dynamite........8vo,	4	00
Fodr's Boiler Making for Boiler Makers................18mo,	1	00
Metcalfe's Cost of Manufactures.........................8vo,	5	00
Metcalf's Steel—A Manual for Steel Users..............12mo,	2	00
*Reisig's Guide to Piece Dyeing..........................8vo,	25	00
Spencer's Sugar Manufacturer's Handbook....16mo, morocco,	2	00
" Handbook for Chemists of Beet Sugar Houses. 16mo, morocco,	3	00
Thurston's Manual of Steam Boilers.......................8vo,	5	00
Walke's Lectures on Explosives...........................8vo,	4	00
West's American Foundry Practice.....................12mo,	2	50
" Moulder's Text-book..........................12mo,	2	50
Wiechmann's Sugar Analysis....................Small 8vo,	2	50
Woodbury's Fire Protection of Mills......................8vo,	2	50

MATERIALS OF ENGINEERING.

STRENGTH—ELASTICITY—RESISTANCE, ETC.

(*See also* ENGINEERING, p. 7.)

Baker's Masonry Construction.............................8vo,	5	00
Beardslee and Kent's Strength of Wrought Iron...........8vo,	1	50
Bovey's Strength of Materials............................8vo,	7	50
Burr's Elasticity and Resistance of Materials..............8vo,	5	00
Byrne's Highway Construction............................8vo,	5	00

Church's Mechanics of Engineering—Solids and Fluids.....8vo,	$6 00
Du Bois's Stresses in Framed Structures.............Small 4to,	10 00
Johnson's Materials of Construction......................8vo,	6 00
Lanza's Applied Mechanics................................8vo,	7 50
Martens's Materials. (Henning.)..........8vo. (*In the press.*)	
Merrill's Stones for Building and Decoration..............8vo,	5 00
Merriman's Mechanics of Materials........................8vo,	4 00
" Strength of Materials........................12mo,	1 00
Patton's Treatise on Foundations.........................8vo,	5 00
Rockwell's Roads and Pavements in France.............12mo,	1 25
Spalding's Roads and Pavements........................12mo,	2 00
Thurston's Materials of Construction.....................8vo,	5 00
" Materials of Engineering..............3 vols., 8vo,	8 00
Vol. I., Non-metallic 8vo,	2 00
Vol. II., Iron and Steel...........................8vo,	3 50
Vol. III., Alloys, Brasses, and Bronzes............8vo,	2 50
Wood's Resistance of Materials...........................8vo,	2 00

MATHEMATICS.

CALCULUS—GEOMETRY—TRIGONOMETRY, ETC.

Baker's Elliptic Functions..............................8vo,	1 50
Ballard's Pyramid Problem..............................8vo,	1 50
Barnard's Pyramid Problem.............................8vo,	1 50
*Bass's Differential Calculus............................12mo,	4 00
Briggs's Plane Analytical Geometry....................12mo,	1 00
Chapman's Theory of Equations.......................12mo,	1 50
Compton's Logarithmic Computations..................12mo,	1 50
Davis's Introduction to the Logic of Algebra.............8vo,	1 50
Halsted's Elements of Geometry..........................8vo,	1 75
" Synthetic Geometry..........................8vo,	1 50
Johnson's Curve Tracing...............................12mo,	1 00
" Differential Equations—Ordinary and Partial.	
Small 8vo,	3 50
" Integral Calculus...........................12mo,	1 50
" " " Unabridged. Small 8vo.	
(*In the press.*)	
" Least Squares............................12mo,	1 50
*Ludlow's Logarithmic and Other Tables. (Bass.)........8vo,	2 00
* " Trigonometry with Tables. (Bass.)............8vo,	3 00
*Mahan's Descriptive Geometry (Stone Cutting) 8vo,	1 50
Merriman and Woodward's Higher Mathematics...........8vo,	5 00
Merriman's Method of Least Squares 8vo,	2 00
Parker's Quadrature of the Circle 8vo,	2 50
Rice and Johnson's Differential and Integral Calculus,	
2 vols. in 1, small 8vo,	2 50

Rice and Johnson's Differential Calculus............Small 8vo,	$3 00
" Abridgment of Differential Calculus.	
Small 8vo,	1 50
Totten's Metrology...8vo,	2 50
Warren's Descriptive Geometry...............2 vols., 8vo,	3 50
" Drafting Instruments......................12mo,	1 25
" Free-hand Drawing.........................12mo,	1 00
" Higher Linear Perspective..................8vo,	3 50
" Linear Perspective.........................12mo,	1 00
" Primary Geometry..........................12mo,	75
" Plane Problems.............................12mo,	1 25
" Problems and Theorems....................8vo,	2 50
" Projection Drawing.........................12mo,	1 50
Wood's Co-ordinate Geometry..........................8vo,	2 00
" Trigonometry..............................12mo,	1 00
Woolf's Descriptive Geometry....................Large 8vo,	3 00

MECHANICS—MACHINERY.

TEXT-BOOKS AND PRACTICAL WORKS.

(*See also* ENGINEERING, p. 7.)

Baldwin's Steam Heating for Buildings................12mo,	2 50
Benjamin's Wrinkles and Recipes.....................12mo,	2 00
Chordal's Letters to Mechanics........................12mo,	2 00
Church's Mechanics of Engineering....................8vo,	6 00
" Notes and Examples in Mechanics...........8vo,	2 00
Crehore's Mechanics of the Girder.....................8vo,	5 00
Cromwell's Belts and Pulleys..........................12mo,	1 50
" Toothed Gearing..........................12mo,	1 50
Compton's First Lessons in Metal Working............12mo,	1 50
Compton and De Groodt's Speed Lathe...............12mo,	1 50
Dana's Elementary Mechanics........................12mo,	1 50
Dingey's Machinery Pattern Making..................12mo,	2 00
Dredge's Trans. Exhibits Building, World Exposition.	
Large 4to, half morocco,	10 00
Du Bois's Mechanics. Vol. I., Kinematics 8vo,	3 50
" " Vol. II., Statics.. ....8vo,	4 00
" " Vol. III., Kinetics................8vo,	3 50
Fitzgerald's Boston Machinist........................18mo,	1 00
Flather's Dynamometers.............................12mo,	2 00
" Rope Driving..............................12mo,	2 00
Hall's Car Lubrication................................12mo,	1 00
Holly's Saw Filing 18mo,	75
Johnson's Theoretical Mechanics. An Elementary Treatise	
(*In the press.*)	
Jones's Machine Design. Part I., Kinematics............8vo,	1 50

Jones's Machine Design. Part II., Strength and Proportion of Machine Parts..........8vo,	$3	00
Lanza's Applied Mechanics..........8vo,	7	50
MacCord's Kinematics..........8vo,	5	00
Merriman's Mechanics of Materials..........8vo,	4	00
Metcalfe's Cost of Manufactures..........8vo,	5	00
*Michie's Analytical Mechanics..........8vo,	4	00
Richards's Compressed Air..........12mo,	1	50
Robinson's Principles of Mechanism..........8vo,	3	00
Smith's Press-working of Metals..........8vo,	3	00
Thurston's Friction and Lost Work..........8vo,	3	00
" The Animal as a Machine..........12mo,	1	00
Warren's Machine Construction..........2 vols., 8vo,	7	50
Weisbach's Hydraulics and Hydraulic Motors. (Du Bois.)..8vo,	5	00
" Mechanics of Engineering. Vol. III., Part I., Sec. I. (Klein.)..........8vo,	5	00
Weisbach's Mechanics of Engineering. Vol. III., Part I., Sec. II. (Klein.)..........8vo,	5	00
Weisbach's Steam Engines. (Du Bois.)..........8vo,	5	00
Wood's Analytical Mechanics..........8vo,	3	00
" Elementary Mechanics..........12mo,	1	25
" " " Supplement and Key.....12mo,	1	25

METALLURGY.

Iron—Gold—Silver—Alloys, Etc.

Allen's Tables for Iron Analysis..........8vo,	3	00
Egleston's Gold and Mercury..........Large 8vo,	7	50
" Metallurgy of Silver..........Large 8vo,	7	50
* Kerl's Metallurgy—Copper and Iron..........8vo,	15	00
* " " Steel, Fuel, etc..........8vo,	15	00
Kunhardt's Ore Dressing in Europe..........8vo,	1	50
Metcalf's Steel—A Manual for Steel Users..........12mo,	2	00
O'Driscoll's Treatment of Gold Ores..........8vo,	2	00
Thurston's Iron and Steel..........8vo,	3	50
" Alloys..........8vo,	2	50
Wilson's Cyanide Processes..........12mo,	1	50

MINERALOGY AND MINING.

Mine Accidents—Ventilation—Ore Dressing, Etc.

Barringer's Minerals of Commercial Value....Oblong morocco,	2	50
Beard's Ventilation of Mines..........12mo,	2	50
Boyd's Resources of South Western Virginia..........8vo,	3	00
" Map of South Western Virginia.....Pocket-book form,	2	00

Brush and Penfield's Determinative Mineralogy. New Ed. 8vo,	$4 00
Chester's Catalogue of Minerals..........................8vo,	1 25
" " " " ....Paper,	50
" Dictionary of the Names of Minerals............8vo,	3 00
Dana's American Localities of Minerals............Large 8vo,	1 00
" Descriptive Mineralogy. (E. S.)....Large half morocco,	12 50
" Mineralogy and Petrography. (J. D.)............12mo,	2 00
" Minerals and How to Study Them. (E. S.).......12mo,	1 50
" Text-book of Mineralogy. (E. S.)...New Edition. 8vo,	4 00
* Drinker's Tunnelling, Explosives, Compounds, and Rock Drills. 4to, half morocco,	25 00
Egleston's Catalogue of Minerals and Synonyms..........8vo,	2 50
Eissler's Explosives—Nitroglycerine and Dynamite........8vo,	4 00
Hussak's Rock-forming Minerals. (Smith.)........Small 8vo,	2 00
Ihlseng's Manual of Mining.. ....8vo,	4 00
Kunhardt's Ore Dressing in Europe......................8vo,	1 50
O'Driscoll's Treatment of Gold Ores.....................8vo,	2 00
* Penfield's Record of Mineral Tests........ ....Paper, 8vo,	50
Rosenbusch's Microscopical Physiography of Minerals and Rocks. (Iddings.)..............................8vo,	5 00
Sawyer's Accidents in Mines......................Large 8vo,	7 00
Stockbridge's Rocks and Soils............................8vo,	2 50
Walke's Lectures on Explosives..........................8vo,	4 00
Williams's Lithology....................................8vo,	3 00
Wilson's Mine Ventilation..............................12mo,	1 25
" Hydraulic and Placer Mining..... ....12mo,	2 50

STEAM AND ELECTRICAL ENGINES, BOILERS, Etc.

STATIONARY—MARINE—LOCOMOTIVE—GAS ENGINES, ETC.

(*See also* ENGINEERING, p. 7.)

Baldwin's Steam Heating for Buildings.................12mo,	2 50
Clerk's Gas Engine....................... ....Small 8vo,	4 00
Ford's Boiler Making for Boiler Makers.................18mo,	1 00
Hemenway's Indicator Practice.........................12mo,	2 00
Hoadley's Warm-blast Furnace..........................8vo,	1 50
Kneass's Practice and Theory of the Injector 8vo,	1 50
MacCord's Slide Valve..................................8vo,	2 00
Meyer's Modern Locomotive Construction.................4to,	10 00
Peabody and Miller's Steam-boilers.....................8vo,	4 00
Peabody's Tables of Saturated Steam....................8vo,	1 00
" Thermodynamics of the Steam Engine......... 8vo,	5 00
" Valve Gears for the Steam Engine.............8vo,	2 50
Pray's Twenty Years with the Indicator............Large 8vo,	2 50
Pupin and Osterberg's Thermodynamics................12mo.	1 25

Reagan's Steam and Electric Locomotives............12mo,	$2	00
Röntgen's Thermodynamics. (Du Bois.)..............8vo,	5	00
Sinclair's Locomotive Running.....................12mo,	2	00
Snow's Steam-boiler Practice........8vo. (*In the press*.)		
Thurston's Boiler Explosions.......................12mo,	1	50
" Engine and Boiler Trials.....................8vo,	5	00
" Manual of the Steam Engine. Part I., Structure and Theory.............................8vo,	6	00
" Manual of the Steam Engine. Part II., Design, Construction, and Operation.............8vo,	6	00
2 parts,	10	00
Thurston's Philosophy of the Steam Engine...........12mo,		75
" Reflection on the Motive Power of Heat. (Carnot.) 12mo,	1	50
" Stationary Steam Engines..................8vo,	2	50
" Steam-boiler Construction and Operation......8vo,	5	00
Spangler's Valve Gears..............................8vo,	2	50
Weisbach's Steam Engine. (Du Bois.)................8vo,	5	00
Whitham's Constructive Steam Engineering............8vo,	6	00
" Steam-engine Design......................8vo,	5	00
Wilson's Steam Boilers. (Flather.)..................12mo,	2	50
Wood's Thermodynamics, Heat Motors, etc.............8vo,	4	00

TABLES, WEIGHTS, AND MEASURES.

FOR ACTUARIES, CHEMISTS, ENGINEERS, MECHANICS—METRIC TABLES, ETC.

Adriance's Laboratory Calculations...................12mo,	1	25
Allen's Tables for Iron Analysis.......................8vo,	3	00
Bixby's Graphical Computing Tables.................Sheet,		25
Compton's Logarithms..............................12mo,	1	50
Crandall's Railway and Earthwork Tables..............8vo,	1	50
Egleston's Weights and Measures.....................18mo,		75
Fisher's Table of Cubic Yards....................Cardboard,		25
Hudson's Excavation Tables. Vol. II................8vo,	1	00
Johnson's Stadia and Earthwork Tables...............8vo,	1	25
Ludlow's Logarithmic and Other Tables. (Bass.).......12mo,	2	00
Totten's Metrology..................................8vo,	2	50

VENTILATION.

STEAM HEATING—HOUSE INSPECTION—MINE VENTILATION.

Baldwin's Steam Heating...........................12mo,	2	50
Beard's Ventilation of Mines........................12mo,	2	50
Carpenter's Heating and Ventilating of Buildings.......8vo,	3	00
Gerhard's Sanitary House Inspection.................12mo,	1	00
Reid's Ventilation of American Dwellings.............12mo,	1	50
Wilson's Mine Ventilation..........................12mo,	1	25

MISCELLANEOUS PUBLICATIONS.

Alcott's Gems, Sentiment, Language............Gilt edges,	$5 00
Bailey's The New Tale of a Tub..........................8vo,	75
Ballard's Solution of the Pyramid Problem..............8vo,	1 50
Barnard's The Metrological System of the Great Pyramid..8vo,	1 50
Davis's Elements of Law................................8vo,	2 00
Emmon's Geological Guide-book of the Rocky Mountains..8vo,	1 50
Ferrel's Treatise on the Winds..........................8vo,	4 00
Haines's Addresses Delivered before the Am. Ry. Assn...12mo.	2 50
Mott's The Fallacy of the Present Theory of Sound..Sq. 16mo,	1 00
Perkins's Cornell University.....................Oblong 4to,	1 50
Ricketts's History of Rensselaer Polytechnic Institute.....8vo,	3 00
Rotherham's The New Testament Critically Emphasized. 12mo,	1 50
" The Emphasized New Test. A new translation. Large 8vo,	2 00
Totten's An Important Question in Metrology............8vo,	2 50
Whitehouse's Lake Mœris............................Paper,	25
*Wiley's Yosemite, Alaska, and Yellowstone............4to,	3 00

HEBREW AND CHALDEE TEXT-BOOKS.
For Schools and Theological Seminaries.

Gesenius's Hebrew and Chaldee Lexicon to Old Testament. (Tregelles.)...................Small 4to, half morocco,	5 00
Green's Elementary Hebrew Grammar..................12mo,	1 25
" Grammar of the Hebrew Language (New Edition).8vo,	3 00
" Hebrew Chrestomathy..........................8vo,	2 00
Letteris's Hebrew Bible (Massoretic Notes in English). 8vo, arabesque,	2 25

MEDICAL.

Bull's Maternal Management in Health and Disease.......12mo,	1 00
Hammarsten's Physiological Chemistry. (Mandel.)........8vo,	4 00
Mott's Composition, Digestibility, and Nutritive Value of Food. Large mounted chart,	1 25
Ruddiman's Incompatibilities in Prescriptions............8vo,	2 00
Steel's Treatise on the Diseases of the Ox.... ....8vo,	6 00
" Treatise on the Diseases of the Dog...............8vo,	3 50
Woodhull's Military Hygiene.......................... 16mo,	1 50
Worcester's Small Hospitals—Establishment and Maintenance, including Atkinson's Suggestions for Hospital Architecture..12mo,	1 25

www.ingramcontent.com/pod-product-compliance
Lightning Source LLC
Chambersburg PA
CBHW031745230426
43669CB00007B/495